INTACT

INTACT

A FIRST-HAND ACCOUNT OF THE D-DAY INVASION
FROM A 5TH RANGERS COMPANY COMMANDER

JOHN C. RAAEN JR.

Major General, U.S. Army (Ret.),
author of the Official After Action Report
of the 5th Ranger Infantry Battalion

REEDY PRESS
St. Louis, Missouri

Reedy Press
PO Box 5131
St. Louis, MO 63139, USA

Library of Congress Control Number: 2012935207

ISBN: 978-1-935806-27-1

Please visit our website at www.reedypress.com.

Cover design by Rob Staggenborg

Printed in the United States of America
12 13 14 15 16   5 4 3 2 1

CONTENTS

Introduction   vii

Map of the Invasion Area   xiii

Maps of Vierville and Pointe du Hoc Areas   xiv

Map of Grandcamp Maisy Area   xvi

Chapter 1 - The Ship - 0001   3

Chapter 2 - The Boats - 0500   9

Chapter 3 - The Plan   14

Chapter 4 - The Approach - 0700   23

Chapter 5 - The Approach (continued) - 0710   33

Chapter 6 - The Beach - 0745   38

Chapter 7 - Assault Up the Bluffs - 0810   51

Chapter 8 - The Hedgerows - 0830   62

Chapter 9 - Battle Along the Coastal Highway   71

Chapter 10 - Vierville-sur-Mer - 1700   79

Chapter 11 - St. Pierre du Mont   84

Chapter 12 - Relief of Pointe du Hoc and the Rest   99

Endnotes   110

References   113

Appendix A - After Action Report, 5th Rangers Battalion   118

Appendix B - The Story of Father Lacy   128

Appendix C - Headquarters Roster   132

Appendix D - Awards - Citations for the DSC   139

Appendix E - Battle Honors   148

Appendix F - Citation for the Silver Star   150

Appendix G - Extract of Orders   152

Appendix H - After Action Report, Motor Launch 304   154

Appendix I - Things Historians Don't Know   158

Appendix J - Photographs   161

Index   171

INTRODUCTION

There were two reasons for writing this book. The first was to get all my memories and notes into a single coherent story. At the same time, and part of point one, I hoped to bring my memories into as close agreement as integrity permits with the stories of others. My personal sources are a letter written on July 8, 1944. The next is a handwritten draft without a date, but which must have been written in July or August of 1944. Most of the story is based on these two papers. Another typewritten paper from 1945 furnishes a bit more information. The rest is memories and the stories of others.

Quite frankly, I had forgotten about the three papers when sometime in 1984 I tried to reconstruct what had happened to me on June 6, 1944. This time I used Henry S. Glassman's story *Lead the Way, Rangers*, the story of the 5th Rangers he took from the official records of the battalion. There were other sources as well that helped a little bit, *The Longest Day*, *Small Unit Actions*, and *Omaha*

Beachhead, to name the most important. Only after I finished my story in 1984 did I find the "lost" letter and manuscripts. What a shock! My memories forty years later just did not reflect the stories I had written contemporaneously with the action.

In 1992, after attending the fiftieth anniversary celebration of the organization of our modern Rangers, I decided it was time to try again.

Originally I had thought to take Robert W. Black's fine book, *Rangers in World War II*, as my "text" and annotate it as I felt necessary. As I proceeded, making marginal notes, I realized that I was commenting on nearly every statement he made. Not necessarily disagreeing but more in the sense of expanding. Since so many of the words were mine, why not write the story from scratch in my own words. As I progressed, I realized my original penciled manuscript written in the summer of 1944 told the story just the way I wanted to tell it. So, quite frankly, I copied almost every word of it, embellishing where necessary from my own memories and those of others.

My own story is autobiographical and told in the first person. I decided to tell this story in normal type with the stories of others and those from other sources in italics, often indented, attributing them where possible to other authors. Many of the italicized stories were as familiar to me as my own, but only through the telling by others. I tried to put it all together as honestly as possible. The times certain events happened were very hard to set at first, and then suddenly all the actions seemed to flow with a common clock.

If you have gotten this far, you should know that the second reason I wrote this story was in the hopes that any member of the

5th Rangers would use it as a framework for his own story. And then, of course, he could send the stories to me to be added to this manuscript.

After I wrote the first version, I circulated it to a few people and received back some of the hoped-for responses. The first was from Jim Graves. Mac McIlwain, Theodore Wells, and Ace Parker followed, as well as others. With each new story input, I had to rewrite much of the basic story. Thank goodness for computers and word-processor programs. I never would have attempted this without them.

Much later, I found myself working with other authors of D-Day stories. Most were kind enough to share some of their sources with me. Joe Balkoski and Kevan Elsby came up with the interview notes of Lieutenant Colonel Charles Taylor, the Army historian who, along with Sergeant Forrest Pogue, interviewed the members of the 2nd and 5th Ranger Infantry Battalions about their D-Day experiences. I used these "War Department Notes," or "WD Notes" extensively since they cover the whole gamut of actions of the two battalions and are necessary in any coverage of the 5th Ranger Infantry Battalion.

I thought I had finished the *Intact* story. True, it would have been nice to embellish the story of D+1 with the information in the After Action Report, but that was hardly worth it.

Then, in March of 2004, I was attending a meeting called by the American Battle Monuments Commission about the new Interpretive Center for the Normandy Cemetery at Colleville, France. One of the attendees was new to me, Lieutenant Colonel Mark Reardon, a senior historian for the Army Office of Military History and a published author in his own right. He said he had several documents I might be interested in. They were in the first mail after I got home. And what documents! After Action Reports from many of the Royal Navy participants in the Invasion. The Royal Navy had destroyed these reports after the war and suddenly they were available again. I sent copies to Kevan Elsby and the British Flotilla Association, and they were overjoyed. Because it is of such great interest to historians, I have included a copy of the report of ML 304, the guide boat for Colonel Rudder's Pointe du Hoc force as Appendix H.

Another bonanza came in 2005 from Dean Thomas Hatfield of the University of Texas. Dean Hatfield was writing a biography of Colonel Rudder. In Colonel Rudder's personal papers he found a copy of Colonel Taylor's War Department Notes of Ranger actions for D+1 and D+2. These notes included our actions at St. Pierre du Mont as well as the 2nd Ranger actions at Pointe du Hoc, which the notes from Elsby and Balkoski had not done. Dean Hatfield was kind enough to send me copies of these notes. They form the basis of much of the St. Pierre du Mont story.

Even more recently, I have been in contact with Gary Sterne, an Englishman who bought the property surrounding the Maisy Battery where the 5th Rangers fought a major battle on D+3. Because of his interest in the history of his property, he has done a great deal of research on the Maisy Battery to include interviewing his French neighbors and many of the Rangers who participated in the battle as well as excavating many of the fortifications on the site. He generously shared the results of his research with me, and it provides much of the detail in Chapter 12.

We all have our heroes. For D-Day, mine were General "Dutch" Cota, Father Joe Lacy, Lieutenant "Ace" Parker, and Captain Bert

Hawks. Each of these men accomplished their assigned missions.

Without General Cota, his heroism and leadership, the Omaha Dog beaches would surely have been lost. True, the 5th Rangers would have gotten off the beach, marched or fought their way to Pointe du Hoc, and relieved the 2nd Rangers. But I doubt that the effort by C Company of the 116th Infantry could have established and held the beachhead without the Rangers there beside them.

Father Lacy was just a plain old ordinary hero. He did not influence the action. All he did was serve the wounded, the dying, and the dead at the water's edge of Omaha Dog Beach. Dragging the wounded out of the surf's advancing edge, comforting them, tending their wounds, putting them in places of relative safety behind the obstacles and wreckage, saying the last rites over the dead—all this completely exposed to the very fire that was killing and wounding those around him. He only had joined us a week or so before the invasion, but he proved he was a Ranger. An expanded story about Father Lacy is found in Appendix B. Father Lacy's citation for the Distinguished Service Cross is included in Appendix D.

Ace Parker, the commanding officer of A Company, 5th Rangers, was the only leader to accomplish the 5th Ranger mission. He reached Pointe du Hoc on D-Day. The rest of us did not make it until D+2. Ace's citation for the Distinguished Service Cross is included in Appendix D.

There was a fourth hero for me, although I did not meet him until D+1. Captain Berthier B. Hawks, C Company, 116th Infantry. Despite a foot crushed during the landing, Captain Hawks was the first to lead his company off Omaha Dog Beach. It was his company that I saw fighting its way up the bluffs to my right just after the 5th Rangers landed and before we started our attack. Hawks was still with us when we started the attack on D+2 that relieved the 2nd Rangers at Pointe du Hoc.

Map coordinates are used extensively throughout the text of *Intact*. To fully appreciate the story and to locate positions on the maps, you should learn how to read coordinates. In the text, the coordinates of a point are shown in brackets as {643918}.

The first step in using coordinates is to divide that six-digit number into equal halves, "643" and "918." The "643" is the "X" or east-west half of the coordinate, the "918" is the "Y" or north-south half. On the chart below, find the line running north-south marked "64." Move 3/10 of the way to the next line running north-south, "65." Mark that spot, usually with a moveable straightedge. On the chart it is marked with a small vertical arrow. Next find the east-west line marked "91." Move 8/10 of the distance to the next east-west line, "92." On the chart, that point is shown by a small horizontal arrow. {643918} is shown on the chart as a star and is the intersection of the lines drawn from those two points. On the Vierville-sur-mer map (Map 3), the coordinates of the Vierville exit are {648918}. The tip of Pointe du Hoc on Map 2, is {586940}. A six-digit coordinate gives the position to within one hundred meters. By adding two more digits you can locate the position to within ten meters. The coordinates of the spot where I landed are {65949104}.

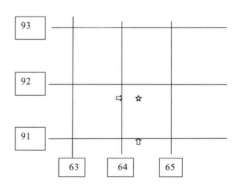

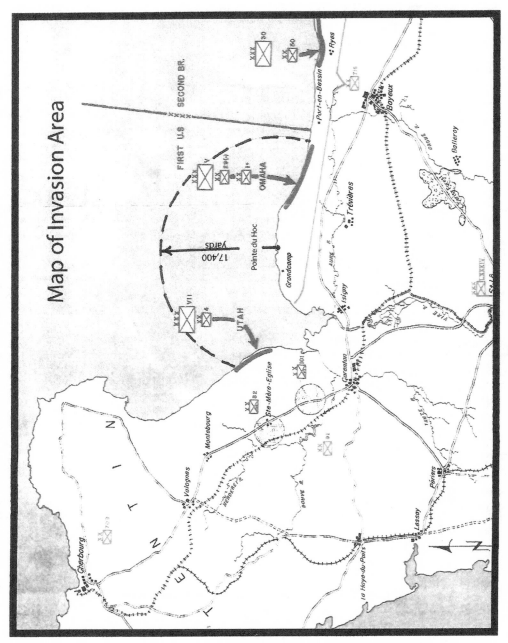

Map 1
The Invasion Area
Coverage of Pointe du Hoc Guns

INTACT   xiii

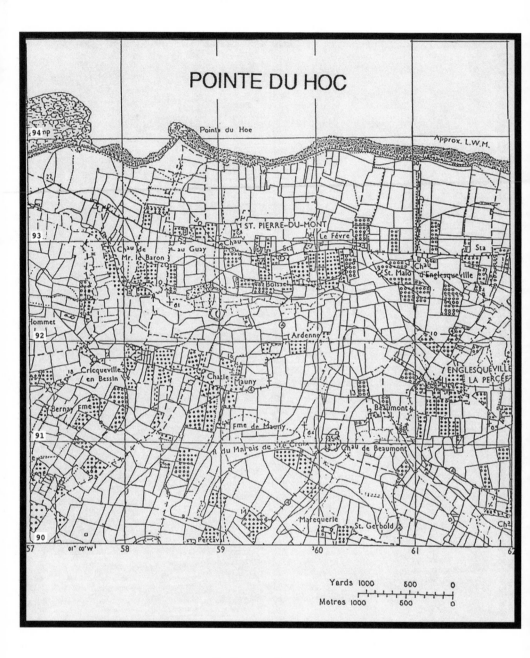

Map 2
Pointe du Hoc

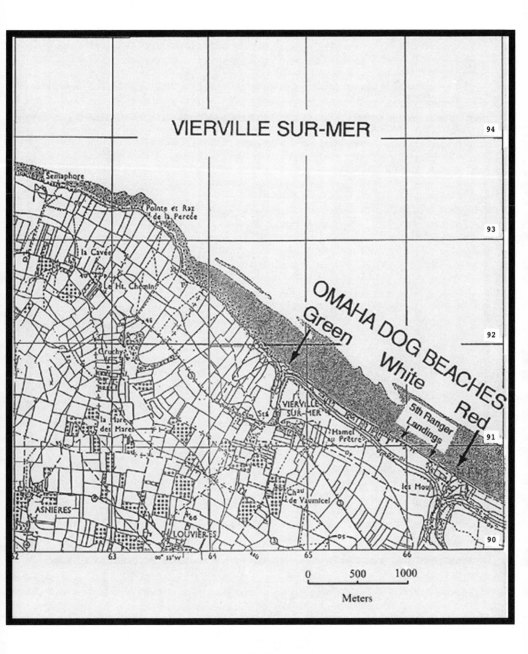

Map 3
Vierville

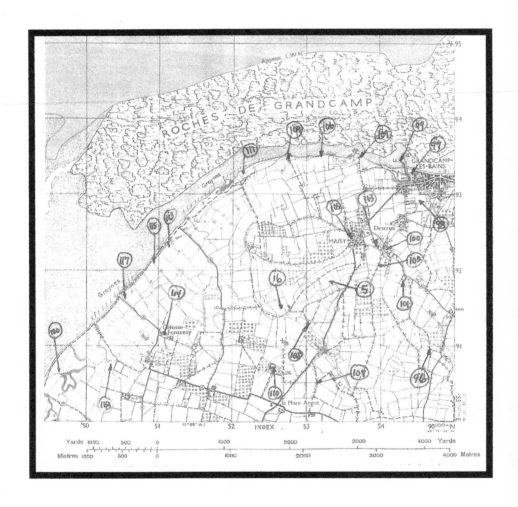

Map 4
Grandcamp-Maisy
With Artillery Registration Points

INTACT

chapter 1

THE SHIP

0001

————————

A t midnight of June 5, 1944, I was on the bridge of the HMS *Prince Baudouin*, an old channel steamer pressed into service by the Royal Navy as a troopship for the invasion of Normandy. The *Prince Baudouin* held one half of the 5th Ranger Infantry Battalion (5th RIB) (Companies C, D, F, and one-half of Headquarters Company) as it sailed toward the coast of Normandy. The other half of the battalion (Companies A, B, E, and the other half of Headquarters Company) was aboard the HMS *Prince Leopold*, another channel steamer.

It was absolutely pitch black on the bridge, cold and stormy. My job was twofold. I was the liaison officer between the ship's captain, Lieutenant Commander W. E. Gelling, RNR, and Major Richard P. Sullivan, USANG, the Ranger troop commander. To assist me in that task, I had a runner. For the second job, I had two Ranger riflemen, one on the starboard wing of the bridge and the other on the port wing. The riflemen were to shoot at and explode any floating mines they or anyone else observed. It was an impossible job. We couldn't see a thing except the fluorescence of an occasional breaking wave, or a deeper shadow that would turn out to be a funnel or a steward with coffee.

My watch had begun at 2200 hours on June 5. At midnight I was scheduled to be relieved, lay below, and get two or three hours' sleep. But who could sleep on a night like this? Toss maybe. Sleep? No! Worry? Yes! Check my gear ten more times? At least. When midnight came, I decided to let my relief sleep a couple of hours more, if he could. After all, he was a good friend, Captain Bill Wise, the C company commander. He was going to be as busy as the proverbial one-armed paper hanger in a few hours. I, on the other hand, was Headquarters company commander, and with our mission, I couldn't visualize setting up a battalion command post in the first few hours after we hit the beach. So sleep on, Bill. I did relieve the runner and riflemen with new men.

At 0200 hours, I was again supposed to be relieved, this time by Major Dick Sullivan, the battalion executive officer and commander of the half battalion aboard the *Prince Baudouin*. Sully, as he was better known, was also the executive officer of the Provisional Ranger Group. This group consisted of the two Ranger battalions, the 2nd and 5th, as well as detachments from the 293rd Joint Assault Signal Company and the 165th Signal Photo Company, all under the command of Lieutenant Colonel James Earl Rudder. Colonel Rudder was also the commanding officer of the 2nd Ranger Battalion. Again I decided to let my relief sleep in, if he could. He too would be a little busy in a few hours.

But now, things were beginning to happen. A huge shadow began to lurk to our port. Gradually it materialized into another ship steaming parallel to us, headed for the Normandy Coast. And then in the gray light of pre-dawn, forms of other ships slowly came into view. In the morning twilight, the horizon took shape and I could begin to see our destination, a dull red glow that was the coast of France. If I strained, I could just barely hear the rumbling of bombs as Allied aircraft battered Normandy yet one last time before the troop landings.

Things were stirring aboard ship as well. At 0334, the anchor was dropped.[1] No more need for me on the bridge. Troops rousted out of sleep. Chow down. Check equipment and put it on.

The standard equipment for a Ranger consisted of helmet and helmet liner, with some wearing the wool knit cap under the helmet. Most stuffed toilet paper into the harness of the helmet liner. The helmets had no covers or netting, but later, after we had salvaged lots of camouflaged parachute silk, many men improvised camouflaged helmet covers. The battle dress was an impregnated, olive drab, fatigue-type uniform. The cloth was stiff and smelled from the impregnation that was designed to reduce chemical warfare casualties should the Germans use toxic chemicals. Field jackets and stripped packs were next, together with the rifle belt, carrying ten clips of eight rounds, and its suspenders. Most of the men were armed with the M-1 Rifle and carried two bandoleers, twelve eight-round clips, of extra rifle ammunition. Two to four grenades were taped to the suspenders. Most of the grenades were fragmentation, but a sprinkling were smoke, concussion, or thermite. A bayonet and entrenching tool were attached to the pack. A poncho and personal articles were carried in the pack. First-aid packs with morphine syrettes, a fighting knife, lensatic compass, and canteen were attached to the belt. A gas mask and one or two inflatable life belts hung around the waist. Three days' worth of D rations were carried in the pack. Many wore wrist watches and carried field glasses and a caliber .45, M1911A1 automatic pistol. Bloused trousers and jump boots completed the array. Cavalry-type blanket rolls with the rest of the Rangers personal equipment were stored with the rear echelon and, if all went as planned, would be delivered on D+1.

Shortly before 0500, it happened. "Attention on deck! Attention on deck! All boat crews boat stations! All boat crews boat stations!" Only a few minutes to go.

"Attention on deck! Attention on deck! United States Rangers, embarkation stations! United States Rangers, embarkation stations!" This was it! Move up to the boat deck. Count your men. Count them again. Check your equipment once more. Check your men's equipment. Now climb into the boats, mine was LCA 1377, and be careful as boats slammed back and forth in the rough sea.

We Rangers were lucky. Our boats were LCAs, Landing Craft Assault. These were Higgins-type boats that looked at lot like the LCPR, Landing Craft Personnel with Ramp, but the LCA was British manufactured. The LCA, like its principal counterpart the LCVP, Landing Craft Vehicular Personnel, carried thirty-three armed and equipped infantrymen. The LCA had a crew of four—skipper, coxswain, seaman/engineer, and seaman/signals. The LCA rode lower in the water than the LCVP and had a more powerful, more silent engine. Luck was that the boats were slung from davits on the boat deck, and we didn't have to climb down cargo netting only to end up jumping into a boat bouncing around in the rough seas.

There was one drawback with the LCA. The exit ramp was too narrow for more than one man to leave the craft at a time. Under the best conditions it took two minutes to clear the boat of men and equipment. Under combat conditions and sea conditions such as we would have in Normandy, it probably took close to five minutes to unload.

Through all this activity aboard the Prince Baudouin, *Sergeant James W. Graves Jr. was having similar experiences aboard the HMS* Prince Leopold, *another Belgian channel steamer holding the other half of the 5th Ranger Battalion. Sergeant Graves was in the Communications Platoon of Headquarters Company. This platoon had been given the duty of maintaining a watch in the command center located directly below the bridge of the* Prince Leopold. *The job here was to make sure that the officers of the 5th Battalion were made aware of everything that happened during the night and to relay orders if there were any change of plans, or any new information that needed to be passed on. The night before, there had been such information, for the landings on June 5 were called off, the armada turned around and returned to port.*

Graves, as communications sergeant, handled the roster for this duty and had arranged it so that he would be on watch when daylight came. The weather on the morning of June 6th was foul with a biting wind blowing up white caps and light fog swirling before the wind. Visibility was about four miles after daybreak.

Lieutenant Colonel Max F. Schneider, commanding officer of the 5th Ranger Infantry Battalion, joined Graves in the command center just before 0500 hours. "Not a very pretty morning is it, Sergeant?"

"No, sir," Graves answered, "it's a very cold morning."

Schneider said to him, "Sergeant, you can go below now and get your equipment."

"Thank you, sir, but I have my equipment. I brought it up with me when I came on duty."

Graves's equipment was critical to Schneider for he was Schneider's radio operator. In addition to the infantryman's tools of the trade, Graves carried an SCR 284 backpack radio weighing thirty-two pounds and a Model 209 Code Converter. To put it mildly, Graves was really loaded down.

At this point, the order for the Rangers to assemble on the boat deck was given and the men began to file up to the deck. From here on, the launching of the boats went much the same as on the Baudouin.[2]

chapter 2

THE BOATS

0500

We were lowered away into the violent sea. Some of the davits didn't work properly and boats almost flipped end for end. Boats smashed into the side of the ship, upsetting everything aboard. Waves lashed at the keels as the LCAs were lowered into the water. Hooks jammed, tackle had to be cut away to get free of the ship, but finally we were clear, seven boats wallowing and plunging in the heavy seas.

First Lieutenant Stanley Askin and his cabin mate First Lieutenant John Reville had been awakened at 0400 by a Ranger from Reville's F Company. Askin had been in A Company but had been switched to the battalion staff for the invasion as an assistant to Major Sullivan. Reville had things to do to get his platoon up and ready to load into the LCAs,

9

so he dressed, grabbed his equipment, and left immediately.

Askin had only to get himself ready, so he decided to grab a few more winks. Big mistake! He suddenly woke up with no idea of how long he had slept. What time was it? Noises? Where were the noises that troops make in the passageways and on the ladders? Silence. He grabbed his equipment and rushed out and up the ladder onto the boat deck. His boat was nowhere in sight and others he could see were being lowered. My God! They've already gone! He rushed to the side of the Baudouin *where he could see LCA 1377 just below him. He jumped, landed safely on the fantail, and slipped into the well of the LCA. The loud thump of his landing was nothing compared to the noise of the boat smashing into the side of the* Baudouin. *Neither Major Sullivan nor I even knew he was late for formation.*[3]

As we circled into a double column to head for our objective, we again heard the loudspeakers blast out. This time it was the *Prince Baudouin*'s captain, W. E. Gelling, wishing us, "Good-bye rangers and God Bless You." It may sound melodramatic now, but at the time, we needed those words of send-off.

Departure from the Baudouin *must have been shortly after 0500, although Robert Black in* Rangers in World War II[4] *and the After Action Report of Lieutenant E. H. West, RNVR, leader of the 507th Assault Flotilla both put the time of our departure at 0545.*[5] *This time cannot be correct, since we were well underway and abreast of the USS* Texas *when it opened the bombardment at 0550 from a range of 13,000 yards.*[6] *The boats of the 504th Assault Flotilla carrying Colonel Schneider*

There were seven LCAs aboard the *Prince Baudouin.* They were from the 507th Assault Flotilla. The boats were numbered: 521, 554, 577, 578, 670, 863, and 1377. Royal Navy records show that none of the boats in the 507th Flotilla were lost, although LCA 1377 was hit by artillery after landing its Rangers.

After circling, Colonel Schneider's boat took the lead in the first wave with the other boats of the 504th Assault Flotilla fanning out to the right and left in the traditional double-column formation of the Royal Navy. Sullivan's boat took the lead in the second wave. Somewhere along the way, we were joined by six LCAs of the 501st Assault Flotilla carrying the Provisional Ranger Group Headquarters, Companies A and B and part of Headquarters of the 2nd Ranger Battalion. The 501st became our leading wave. We were now three waves of landing craft.

"The run in towards Dog Green Beach was at approximately 5 knots. Difficulty was experienced in keeping station owing to the condition of sea and swell. But overall, the craft were able to keep fairly good formation."[8]

For ourselves, we were crammed into three rows, no shoulder room, no knee room (see photo on page 166). The LCA rides low in the water, so even the smallest wave showered us. And the sea was rough. Waves of six to eight feet battered the tiny transports. Drenched and miserable, some of the men began to get seasick. But not to worry, we had only two hours more to go.

Before leaving the ship, each man was issued three waxed paper sacks. If for some strange reason you became seasick, you were supposed to use these bags and throw them overboard. For those who were lucky enough to avoid becoming seasick, their waxed bags, affectionately known as "puke bags," were borrowed. But soon even these were gone and the boats began to reek of vomit.

Major Sullivan ran a well-disciplined boat, so only he, and occasionally I, stood up to see our progress. Sunrise was at 0558. As I remember it, we were on British double daylight time. Morning astronomical twilight, better known as the crack of dawn, occurred at 0446. So many boats, so many ships, at first, but then we were clear of the marshaling area, and we almost seemed alone in the sea. At a great distance to our right, I could make out warships maneuvering. Much of the coast to our south (front) was burning with smoke drifting to the left, but mostly I stayed down, more to set an example for the men than anything else. The Royal Navy officer-in-charge and the coxswain were standing behind the moderate protection of some steel plate.

Suddenly there was a tremendous crash, roar, blast. Sully and I jumped up, but the officer-in-charge, Sub-Lieutenant Ernest Pallent, RNVR, calmly said, "Sirs, that is the Battleship *Texas* opening the bombardment of the coast." The time was 0550, forty minutes before H-Hour. Now every warship in all directions opened fire with hellacious noise and concussion. *(Max Hastings in* Overlord[9] *says 9 battleships, 23 cruisers, 104 destroyers, and 71 corvettes for the entire invasion.* Omaha Beachhead *says the U.S. Naval Forces in the bombardment were 2 battleships, 3 cruisers, 8 destroyers, as well as numerous other landing craft with one type or another artillery aboard.)*

As for Sergeant Graves, he was watching the Texas *when he was dismayed to see her suddenly engulfed in a great cloud of brown, yellow, and white smoke. His first thought was, "Jesus! The Texas has blown up!" He quickly realized that she had fired her big guns. The sound ripped across the water as a great roar and he turned so that he would be able to see the shells hit the coast of France, just over the horizon. He was able to see large flashes of light about the time he estimated the shells would hit the coast.*[10]

Many of the Rangers' LCAs had trouble with the high seas. The amount of water taken aboard seemed to be dependent on speed. Speed up, take on more water. Slow down, fall behind the other boats in the formation but take on less water. Fortunately, there were delays at various control points and the boats that had fallen behind were able to catch up. Most of the Ranger boats had some trouble with taking on water.[11]

Shortly after this, perhaps five miles from the beach, one of F Company's boats, LCA 578, containing Lieutenant William M. Runge, company commander, and half of F Company headquarters, and Lieutenant John J. Reville and his 1st Platoon of F Company, began to ship more water than the pumps or the men bailing with their helmets could handle. The rest of our boats hesitated, but we too were wallowing in the heavy seas, overloaded, taking about as much water as we could handle. We had to abandon the sinking LCA as other larger craft came to their rescue. We heard later an LCT (*LCT 88*)[12] had taken them aboard and put them ashore on Omaha Easy Green Beach near St. Laurent at 0900 on D-Day. *LCA 578 did not sink, but was taken under tow until her engines could be restarted. After intensive bailing, she was able to return to the* Baudouin *under her own power.*[13]

chapter 3

THE PLAN

The Allied plan for the invasion of Europe was for the U.S. Forces to land in the Normandy area (see Map 1). Two beaches were selected. Utah, lying along the east coast of the Cotentin Peninsula and to its east, Omaha, lying on the north coast of Normandy between the Vire Estuary and Port-en-Bessin. Utah would be assaulted by the U.S. 4th Infantry Division, and Omaha by the 1st Infantry Division with elements of the 29th Infantry Division attached for the assault itself. Omaha Beach was further divided into three beaches, Dog on the west or right, Easy in the center, and Fox on the left (east). Long before the infantry landed on the beaches, the 82nd and 101st Airborne Divisions would land by parachute and glider behind the Utah beaches.

British and Canadian forces would land further to the east of the Americans on beaches known as Gold, Sword, and Juno.

The assaulting troops would disembark from ships about twelve to fifteen miles from Utah and Omaha beaches. Mostly in small boats like LCAs, the assault troops would sail to their respective beaches. Aerial bombardment and naval gunfire would reduce the shore defenses, or at least occupy them until the troops got ashore. The beachhead would then be expanded inland. There was just one little catch: Pointe du Hoc.

Pointe du Hoc ("Hoc" is pronounced "awk" in French) lies midway between Utah and Omaha beaches. It is like a dagger pointing out into the troopship anchorage area of the invasion fleet. Equally bad, it overlooked both Utah and Omaha beaches and the boat routes in from the ships. The Germans had fortified Pointe du Hoc with six captured French 155-mm, GPF, World War I cannon. Two of the gun positions were casemated, the others dug in below ground (see photo on page 161). In all probability, those guns would be equipped with World War I ammunition. The range of this ammunition was 17,400 yards (see Map 1). That meant the guns could completely cover both Omaha and Utah beaches. Further, they could fire on the assault boats as they came in from the anchorage or marshaling area. The marshaling area lay from twelve to fifteen miles out, beyond the ten-mile range of the guns at Pointe du Hoc. But the guns were more frightening than that. The U.S. Army Ordnance Department had achieved a range of 25,000 yards in those guns using modernized ammunition and modified carriages that allowed the guns to elevate beyond thirty-five degrees. If the Germans were going to use modernized guns and ammunition, the 25,000-yard range of the guns easily covered the entire troopship anchorage area as well as both Utah and Omaha beaches.

The Pointe du Hoc artillery pieces had to be neutralized, which became the job of the 2nd and 5th Ranger Battalions. These two Ranger Battalions were formed into a Provisional Ranger Group, commanded by Lieutenant Colonel James Earl Rudder, the 2nd Ranger Battalion commander. For his Group Headquarters, Colonel Rudder borrowed most of his officers and men from the Headquarters of the 2nd Ranger Battalion, although two officers and eleven enlisted men came from Headquarters Company of the 5th Battalion. Major Sullivan, for example, doubled as the group executive officer and 5th Ranger Infantry Battalion executive officer.

Under the plan as finally developed, Pointe du Hoc would be assaulted by three companies, D, E, and F from the 2nd Ranger Battalion. Part of the 2nd Rangers Headquarters would accompany this assault. This was known as Force A. The 100- to 120-foot cliffs of Pointe du Hoc would have to be climbed, even though the enemy was entrenched above.

Meanwhile, the 5th Rangers—with A and B Companies of the 2nd Ranger Battalion, Provisional Group Headquarters, and part of the headquarters of the 2nd Rangers attached—would land on Omaha Beach at the Vierville exit behind the 1st Battalion of the 116th Infantry. These Ranger units comprised Force C. Force C would infiltrate through the assaulting infantry units of the 116th Infantry Regiment. The Ranger companies would assemble after the infiltration at designated rallying points, reorganize, and move to the battalion rendezvous point, a crossroads about one mile to the southwest of Vierville. From there, with A and B Companies of the 2nd Rangers providing the advance guard and flank protection, the 5th Ranger Battalion would advance as rapidly as possible to Pointe du Hoc, attacking the German positions from the land side.

Before any advance on Pointe du Hoc could be undertaken by Force C, there were difficult obstacles that would have to be overcome. Omaha Beach stretched four miles from one end to the other. The beach was backed by a flat table, or plateau, that stretched from a few yards to perhaps a hundred yards. Beyond that level area were steep bluffs rising, in most places, over 100 feet. Beyond the bluffs was the fairly level bocage, or hedgerows and farmland country. Leading up from the beaches and through the bluffs, were five ravines that the Allies expected to use as exits for their tanks and supply trucks. These exits had roads in most cases, but mere paths in others.

In the water in front of the beaches the Germans had emplaced row upon row of obstacles designed to prevent boats from reaching the shore or, at worst, to so disorganize the landing craft formations that any troops reaching the shore would be scattered and disorganized.

On the bluffs behind the beaches, the Germans had built fourteen strongpoints, or *widerstandsnester* (WN), as they called them. Each was heavily protected by minefields, barbed wire, and ditches. Each was heavily armed with artillery, mortars, machine guns, and a small infantry defense force. Between these strongpoints were trenches, machine gun and mortar emplacements, and observation points manned by other infantry units. The two WNs at the opposite ends of the beach each had the dreaded 88-mm anti-aircraft/anti-tank guns firing directly down the beach.

The WNs were concentrated at the ravines or exits from the beach, but a few were located between the exits. The concept was that each WN concentrated its fires to protect the front of adjacent WNs and the defensive positions in between WNs. The infantry force in a WN was too small to do much in the WN's own defense.

WNs 71, 72, and 73 guarded the Vierville exit. Each had weapons that fired directly down the beach, protecting each other as well as delivering crushing enfilade fires on landing craft and troops debouching from landing craft. Other exits had similar multiple strongpoints guarding the exit ravines.

Between those exits were other, smaller WNs supporting the entrenched infantry on either side of them as well as the fronts and flanks of the more distant exit WNs. One such was WN 70, located at vic {65359130}. WN 70 covered the beach area where Force C would eventually land.

From intelligence reports, the Allies knew that the Germans had some six hundred men manning the WNs and entrenchments with an infantry battalion located inland for counterattacks. About two hundred of these six hundred were in the WNs themselves. What the Allies didn't know was that just before the invasion the Germans had reinforced with another five hundred battle-trained troops to man the entrenchments between the WNs and five more infantry battalions for a counterattacking force.[14]

A Ranger Company had only three officers and sixty-five enlisted men. The company was broken into two platoons of one officer and thirty enlisted men each. Each platoon had two sections consisting of a section leader, five riflemen, and five machine gun crewmen. There was also a weapons section of six men in each platoon to operate a 60-mm mortar.

There were six Ranger companies in each Ranger battalion.

The Ranger Headquarters Company had nine officers and ninety-nine men. Headquarters included the staff sections (six officers), attached medics (one doctor and eleven enlisted men, six of whom were attached to the Ranger companies), a communications platoon, a motor platoon, a mess platoon, and a large supply section.

For the invasion, an overstrength was authorized to cover initial casualties. The 5th Rangers used most of this overstrength to create an 81-mm mortar platoon in C Company and a 60-mm mortar platoon in F Company. C Company's 81-mm mortar platoon was made up to form three mortar crews. Each crew had five Rangers, one carrying the baseplate, one the bipod and one the tube, plus two ammunition bearers each carrying six rounds. Riflemen in the company also carried three or four rounds each. In the 60-mm mortar crew, the gunner carried the assembled weapon, tube bipod, and base plate, with the other four Rangers as ammunition bearers.[15]

The story of the 2nd Rangers attack of Pointe du Hoc is told in *Small Unit Actions*,[16] a War Department publication available through the Government Printing Office. The story includes the detailed plan as well as the combat operations involved. The performance of the 2nd Rangers at Pointe du Hoc was magnificent, but it is not a part of my story. However, the story of Theodore Wells, a member of the 5th Rangers is part of my story. Wells was one of Colonel Rudder's radio operators and took part in the assault of Pointe du Hoc.

There was another point on the Normandy Coast, the Pointe et Raz de la Percée. This fortified position lay beyond the west end of Omaha Beach, about fifteen hundred yards from the Vierville exit. Its weapons raked the Omaha Dog beaches with enfilading automatic small arms and artillery fire. Neutralization of Pointe et Raz de la Percée was assigned to Company C of the 2nd Rangers, also known as Force B. The heroism of these Rangers is beyond description, but it too is not part of my story. The story of this action is interwoven into *Omaha Beachhead*,[17] another War Department publication, also available through the Government Printing Office.

During the Fabius exercise held in May 1944, just before the invasion, someone in the Rangers—who, I don't know—came up with a change to the invasion plans. If the assault on Pointe du Hoc was successful, the 2nd Rangers on Pointe du Hoc would radio the Fifth Ranger Battalion the message "Praise the Lord." The 5th Rangers and A and B Companies of the 2nd Rangers would then land at the base of the cliffs at Pointe du Hoc, scale them, pass through the 2nd Rangers, and then move inland to seize the Vierville-Grandcamp (coastal) road, thereby avoiding the dangerous five-mile advance through enemy territory from Vierville. (The full name of this hamlet was Vierville-sur-Mer.) The entire Ranger force would then attack toward Grandcamp and Isigny, the 29th Division's main objectives.

If this "mission successful" message were not received by H+30 minutes or if the "mission failure" message of "Tilt" were received, Force C, the 5th Battalion and Companies A and B of the 2nd Rangers, would alter course and land on Omaha Dog Green Beach at the Vierville exit (known as D-1) as in the original plan. The Fabius exercise had another change. During the exercise, many units got intermingled. As a result, some troops ended up following the wrong leaders with potentially disastrous results.

To prevent such mixups, each Ranger had a horizontal, four-inch orange diamond painted on the rear of his helmet with a black "5" inside the diamond for the 5th Battalion and a "2" for the 2nd. These diamonds helped immensely in keeping unit integrity under stressful conditions. Non-coms also wore a horizontal one-by-three-inch white stripe on the back of their helmets, while officers wore a similar vertical white stripe.

H-Hour for Omaha Beach was 0630. Strangely enough, on Dog Green Beach, no one was scheduled to land at H-Hour. Instead, the DD (Dual Drive) tanks of the 743rd Tank Battalion were to land at H minus 5 minutes, the first wave of infantry at H+1 minute, the next, a wave of engineers, at H+3, and so on. Because of tidal differences, other beaches had different H-Hours. The British beaches had 0700 and 0730 for H-Hour. On June 5, when landings were originally scheduled, H-Hour for Omaha was at 0610.[18]

The tidal range on the Omaha beaches was eighteen feet at the time of the assault. This tidal range would expose about three hundred yards of firm sand at low tide. The Germans had placed rows of "underwater" obstacles beginning about midway in this tidal flat. The landings were planned for a rising tide, about half flood so that most of the obstacles would be above water and not interfere with the initial assault craft. It also meant that the early assault troops would have to cross 150 yards of beach swept by enemy small arms, mortar, and artillery fire. At half flood, the tide would be advancing about a yard a minute up the beach. Engineer demolition units arriving in those first waves could then destroy the obstacles before the tide covered them, making it easier for the later waves of assault craft to land. First low tide for Omaha Beach on June 6 was at 0525 with first high water at 1100.[19]

Because of a problem that arose after we boarded the ships, Colonel Rudder had to make some quick changes in the command structure of the Ranger force. Instead of landing with the main Ranger force, Force C, on Omaha as planned, Rudder found it necessary to land with the three companies of the 2nd Rangers assaulting Pointe du Hoc. He and his radio operator, Wells and four other Group Headquarters officers and men, literally switched boats as we loaded out for the assault. That switch in plans also meant that Lieutenant Colonel Max Schneider, Commanding Officer of the 5th Ranger Battalion, would command the main Ranger force, Force C.

chapter 4

THE APPROACH

0700

Ranger Force C's three waves, Assault Flotillas 501, 504, and 507, sailed a course midway between Omaha Dog Green Beach and Pointe du Hoc. My recollection is that we slowed down at H+15 while waiting for news from the 2nd Rangers at Pointe du Hoc. At that time, we lay offshore a little to the west of Pointe et Raz de la Percée. By 0700 hours, or H+30, we were frantic. No message of any sort had been received from the 2nd. Over our SCR 300 radio, we had heard a call from a beachmaster saying, in effect, that "Omaha Dog White is clear of enemy. Assault forces meeting no resistance." Finally we heard a feeble radio message that could have been from the 2nd Rangers. It was almost unintelligible, but it did contain a word that sounded like "Charlie." We weren't sure what it meant, but it clearly did not mean success.

The reason was simple. The two pilot boats that were to guide Force A to Pointe du Hoc were damaged and unable to do so. A third boat, ML 304, was pressed into service. Shortly after starting for the shore, both ML 304's radar and loran broke down, forcing the officer-in-charge, Lieutenant C. Beevor, RNVR, to continue navigating by dead reckoning. Because of a lack of familiarity with the coastal features, tidal set, and poor visibility caused by smoke from the naval shelling, the storm, and the early hour, the skipper became confused and changed course toward Pointe et Raz de la Percée, turning away from Pointe du Hoc![20] By the time Colonel Rudder realized this error in navigation, he was nearly to the shore, had to demand a sharp right angle turn, and sail down the coastline. That error in navigation cost him a little over half an hour. Instead of landing at 0630 as planned, the 2nd Rangers landed at Pointe du Hoc at 0708.[21]

Worse, D Company, which was to land on the western cliffs, was forced to land abreast of E and F Companies on the eastern cliffs. Instead of the coordinated attack that had been so carefully planned, the attack degenerated into a piecemeal effort. A few minutes after the touchdown, someone in the 2nd is reported to have radioed "Praise the Lord," but it was too late. Lieutenant Colonel Schneider had been forced to commit Force C to a landing at Vierville. At 0710, the 5th Rangers supposedly received the "Tilt" message.[22]

This looks like the perfect place to tell the story of T/5 Theodore H. Wells, Headquarters Company, 5th Ranger Infantry Battalion.

WELLS AND POINTE DU HOC[23]

"Wells was assigned to the Provisional Group Headquarters as a radio operator, and his job was to contact the 5th Rangers, keep in contact, and report the progress of the 2nd Rangers at Pointe du Hoc to the 5th Rangers. Wells had no 'buddies' in the Provisional Group Headquarters. In fact, only two or three men even knew he was from the 5th Rangers; rather, they thought he was transferred from one of the 2nd's line companies or was one of the recently assigned overstrength soldiers.

"Colonel Rudder was to go in with the Provisional Ranger Group Headquarters, and so was Wells. On June 5, 1944, the Headquarters loaded aboard the HMS Amsterdam *or the* Ben Machree, *Wells never knew which. Aboard, after loading, the Group Headquarters Rangers stayed more or less together. Around 2400 hours, Wells checked his equipment for the last time and turned in. The ships were in heavy seas, and were bouncing all over the place, but he was so exhausted that he fell asleep at once. The next thing he knew, the loud speaker was calling the Rangers to wake up. He had not undressed when he lay down so was ready to move out as soon as he awoke. He looked at his watch. It was 0400 hours, June 6, 1944. His first thought of the day was, 'Theodore, you'll soon be in France.'*

"Wells gathered up his gear, standard for a Ranger, except he carried a caliber .45 instead of a rifle, and an SCR 300 radio that weighed about 50 pounds. His radio was the most important item. He had spent nearly one whole day waterproofing it.

"Like many, he did not eat breakfast that morning. He was already beginning to get seasick and wanted off that ship and onto dry land. Wells boarded LCA 888 with Colonel Rudder and sat in the rear, directly under the coxswain on the left side. Colonel Rudder was on the right side in front, and a staff sergeant was on the left side in the front of the landing craft. Wells did not realize that plans had changed during the night and he and Colonel Rudder were now headed for Pointe du Hoc with the Force A assault companies of the 2nd Rangers. There were fifteen men from Company E in the boat together with six officers and men from 2nd Ranger Headquarters.

"The channel was rough and the waves were high. As Wells looked toward the coast, some four or five miles away, he saw a most awesome sight. There was a huge boiling cloud along the coast several miles ahead, and in the back of the cloud, a red glow shown through. It looked like the whole continent of Europe was on fire. Planes roared overhead and the warships all around them were firing. Wells could see the red blast as the bombs and shells hit their targets. A ship or plane must have laid down a huge smoke screen. As landing craft passed through that smoke, the firing of the ships stopped and the planes stopped their bombing. All was quiet now except the waves slapping the side of the boat.

"Wells was sitting nearest to the coxswain and talked to him on the way in. He learned that the coxswain had been on other landings in Africa. The coxswain was standing in his compartment with a good view of what was happening. Every few minutes Wells would stand up and look about quickly and then sit down again. He remembers that the LCA *changed directions, but he didn't know why. This change in direction was probably the one made when Colonel Rudder discovered that his boats were headed for Pointe et Raz de la Percée rather than Pointe du Hoc.*

"Someone said, 'They're shooting at us!' and everyone ducked down a little lower. Wells couldn't see the enemy and hoped they couldn't see his boat. Shortly thereafter, the coxswain said, 'Look there's the point!' Then the LCA grounded. The ramp dropped, and Colonel Rudder and the staff sergeant went out first. The boat was bouncing about, and it was hard to stand up. Then the coxswain said, 'Men, if you value your lives, you will leave this boat now.' Wells didn't wait for those in front to move, and was far from being the last man out.

"When he jumped from the boat, he landed in water over his head. His feet didn't touch the bottom so he took a deep breath and went under. He took the radio from his back and tried to hold it above the water, but it was too deep. When he did touch bottom, he pushed hard and jumped for the surface. After swimming three or four strokes, he reached the dirt and gravel beach. The beach was only four or five feet wide. Wells had jumped into a bomb or shell crater when he left the landing craft. These huge underwater holes and beach craters with their slick clay sides, made the early going for the Rangers very difficult.

"The beach to his left, some ten to fifteen feet away, was much wider, and a shallow cave on this wide spot was their first CP. Wells saw a staff sergeant digging steps into the cliff where it had been bombed. Although craters had pockmarked

the top and face, it was still a very high cliff. Colonel Rudder was behind the sergeant helping him out. Wells now turned his attention to his pistol and radio. He cleaned and dried the gun, then wiped the radio off as best he could and tried to contact the 5th Rangers. He got no answer. He decided the cliffs were creating interference so he would have to go to higher ground.

"By this time, the CP had moved to the top of the cliff and had been set up in a big shell hole or bomb crater. There were big chunks of concrete lying around, so there must have been a coastal gun emplacement there before the bomb destroyed it. Wells went up the cliff in the steps or foot holes that the staff sergeant had dug. He tried the radio again and again, but got no answer. Trying another tack, he then took all the waterproofing off and disassembled it as much as he could. With no tools, there wasn't much that could be done. He dried it some more, checked the batteries again, put it together, and tried to call the 5th Battalion. Again, no answer. It was just dead and he knew it. He wished he had brought more batteries. Wells's problem with the radio was not unique. Most of the radios that made it up the cliffs had problems of one kind or another, and none of the other radios carried by the 2nd Rangers were able to reach the 5th Rangers either.

"By this time, wounded Rangers were coming in and a few prisoners had been captured. Wells was told by a captain to search the prisoners. He did and took everything but their IDs. If they had any pictures of their families, Wells put them into the coat pocket of their uniform when no one was looking. He worried about this and wondered if he should do it. He *decided that if he were taken prisoner, he would want them to do for him. Each time he did this for a prisoner, the prisoner would lean over and whisper, 'Danke schoen.'*

"*The first wounded man that Wells saw was a Ranger who took a bullet in the side of his helmet. The bullet turned and went between his helmet and his helmet liner. It then went through his shoulder and took with it the metal clasp from his helmet. The wounded were taken down the cliff to the water's edge.*

"*By now, it was about 1200 hours, and he tried again to contact the 5th Battalion. Still no answer. Hoping to hear someone calling, he kept his radio on all the time.*

"*Sometime between 1200 and 1400 hours, a big boat came in to take the wounded to the Battleship Texas. The boat had no engine and the sailors in the crew had to row it in. Wells thought it looked like the whale boats he had seen in the movies. The sea was so rough and the waves were so high that after loading the wounded aboard, the sailors could not launch the boat from shore. After trying several times to get clear of the beach, the sailors stayed on the beach with the Rangers and guarded the prisoners.*

"*Wells had not seen Colonel Rudder since moving to the new CP on top of the cliffs. He finally gave up on the radio and told the captain it was no use wasting any more time on it. He was told to go find Colonel Rudder, since Rudder had not been seen nor heard from on the telephone for some time. Wells asked, 'Which way should I look?' The captain's answer was, 'I don't know, just look out there.' He pointed away from the cliffs. Wells checked his .45, buckled the strap on his hel-*

met, climbed out of the bomb crater, and took off. He thought to himself, 'Now you'll see what real combat is like.'

"Looking everywhere, he moved out very cautiously. 'Out there' it looked like he was on the moon. There were craters everywhere, some small and some large. The large ones were monster holes. He never realized a bomb or shell could make a hole that large. There was no one in sight, so he turned to his right and moved along the cliff edge for a while, then, a few hundred feet in front, he saw a concrete structure and a soldier looking at him. Since he couldn't tell what uniform the man wore, he moved in a little closer. He was able to use several small craters for protection. Dodging and ducking, he never knew if he had been fired on, but finally was able to work close enough to recognize the uniform. It was an American, his rifle pointed in Wells's direction. He jumped into a crater and thought, 'If I can recognize him, he can recognize me, but will he?'

"Both Ranger battalions had an orange diamond painted on the back of each helmet for recognition from the rear. Wells took his helmet off, and while still in the crater, held it up showing the orange diamond. Hearing no shots, he looked over the rim of the crater and saw he was being waved in. He felt a little foolish for doing that, but better safe than sorry. As Wells approached the structure, he recognized that it was a fortification. It had a narrow slit facing the channel so a gun could sweep the cliff, the beaches below, and a large area of the channel.

"When he reached the gun emplacement, there was Colonel Rudder and some of his staff. There were about eight men including Wells. He told Rudder about the radio and that he was sent out to find him since there had not been any communication from him in quite a while. Wells noticed Rudder was wounded in the leg. Colonel Rudder asked Wells if he knew how to repair a telephone cable and Wells said, 'Yes.' He found the cable had been damaged by mortar or artillery fire. He made three splices in the cable from the gun emplacement back to the shell crater at the cliff's edge. It was not easy to splice cable under fire, without the proper tools, but he managed with just a knife.

"After repairing the cable, he was told to go back to the command post and get some ammunition for the staff. A little while later, Colonel Rudder asked him to go back again and bring up some food. It was lonely out there running those errands by himself. When he got back with the food, all of them ate except the man standing on guard in front of the fortification.

"The emplacement was quite spacious inside, built of concrete, with a corridor through the center that led to a back exit, passing three rooms on each side. The gun, which looked like an 88-mm, was blown off its pedestal. On top of the gun was the first dead German soldier of many that Wells was to see during the war.

"When they finished eating, Colonel Rudder told him to take the guard's place outside and keep watch. He stepped outside and stood with his back to the front wall. To his left, he could see a ship shelling a small village, and that was some sight to see. A captain, whose name was Harwood (Jonathan H.), stepped up and stood next to him, watching the shelling *through binoculars. The time was about 1600 hours when suddenly all hell broke loose. Wells heard a tremendous explosion and then nothing.*

"*The next thing he knew, he was being carried back to the fortification. One of the Allied ships had fired on the Ranger position. He was told later that it was the British cruiser Glasgow. Wells survived the explosion because Captain Harwood shielded him from the blast. Harwood took most of the fragments from the shell. He died some time during the night. Wells was blown about twenty feet from where he stood guard. He was numb from the blast. When his feelings came back, he was hurting so badly in his back and legs that he thought he could not stand the pain. Dr. Block (Captain Walter E. Block, MC), who was the 2nd Ranger Battalion medical officer, was with the command post. He had to give Wells two shots of morphine to ease the pain. When the shell burst, it covered Wells with a yellow powder that stained his skin for a week or more. It left a very bitter taste in his mouth as well.*

"*Wells was taken back to the cliff's edge and placed in a bunk bed in a German bunker. A little while later, someone came in and said, 'Take a drink of this.' Wells asked what it was and he said, 'Just drink and don't ask questions.' So he did, and to his surprise it was smooth Scotch whisky. It was whisky liberated from the Germans. Thus ended an extremely long day for Wells. In spite of the pain from his injuries, he went right to sleep.*

"*The next day, D+1, all the wounded were evacuated to the* Battleship Texas."

chapter 5

THE APPROACH

(CONTINUED)

0710

Lieutenant Colonel Schneider, the commanding officer of Ranger Force C and the 5th Ranger Battalion, had waited as long as he could. He had delayed past 0700, but finally, about 0710, he gave in to his very specific orders and committed the battalion and Force C to land at Vierville. It was obvious by now that we would really have to hurry to land on time.

Omaha Dog Beach was divided into three sub-beaches, Dog Green was on the right (west) extending from the Vierville exit to the east 970 yards. For most of the length of Dog Green Beach, the Vierville seawall was the inland edge of the beach. Dog White was in the middle extending another 700 yards to the east. There was no seawall beyond Dog White Beach, just dunes. Dog Red Beach was on the left end of Dog Beach, a short 480 yards long ending at the Les Moulins exit. A small wooden seawall edged the beach. Beyond the Les Moulins exit and to the east, was Easy Green Beach.

The first wave of infantry to land on Omaha Dog Green Beach was to be Company A of the 116th Infantry Regiment. Although the 116th Regiment was part of the 29th Infantry Division, the 116th was attached to the 1st Infantry Division for the actual assault of Normandy. Company A, 116th Infantry landed as planned at the Vierville exit. Company C, 2nd Ranger Infantry Battalion (Force B) landed almost simultaneously to A Company's right. While A Company of the 116th Infantry had landed nearly on schedule at H+6 minutes, half of the tanks that were to land ahead of them did not make it to shore. Within fifteen minutes, all of A Company's officers and most of its non-coms were killed or wounded, putting A Company out of action. Of the 197 infantrymen who landed in that company, 96 percent (189) were casualties on D-Day, 91 of them killed, 98 wounded, and only 8 unscathed. B Company of the 116th fared a little better, but only because several of its boats pulled back and diverted to other nearby beaches. One Company A officer, Lieutenant Nance, the company executive officer, came in with the second wave. He survived.

The LCTs carrying half of the 16 DD tanks of B Company, 743rd Tank Battalion had been sunk by artillery fire coming in to Dog Green Beach. Only eight tanks and one lieutenant survived the landing. These tanks took up positions at the water's edge and started to fire at the enemy fortifications. Companies A and C of the 743rd landed in good order far-

ther to the east. Except for these eight tanks of Company B, the infantry soldier was alone on Omaha Dog Green Beach.

The DD, dual drive, tank was a standard M-4 tank that had been modified to float and propel itself in the water. This was accomplished by placing a huge "inner tube" around the tank, giving it a propeller, and providing it with high air intakes and exhausts to allow the engines to keep running when nearly submerged. The DD tanks were to be launched into the water from LCTs when 6,000 yards from the beach. The 741st Tank Battalion, which landed in support of the 1st Division, tried this, but all but two tanks launched swamped almost immediately. The 743rd's DD tanks were carried into the beach by the LCTs, thanks to the heroism of the LCT crews.[24]

Victor Hicken, a U.S. Navy ensign, was aboard LCT(A) 2227. This craft landed at H-Hour (0630) and unloaded two demolition teams and three DD tanks of B Company of the 743rd just east of the Vierville exit. Hicken watched as Company A of the 116th landed behind and to his right. LCT 2227 became stuck on the beach and was riddled by small arms and artillery fire. Finally, floated by the incoming tide, the LCT backed off the beach at about 0700. As it sailed by a picket boat just after 0700, it was hailed and asked about casualties. The crew reported what had occurred at Vierville. With that the picket boat hailed others in the area and said that Dog Green was closed to further landings.[25]

As Force C approached Vierville, the picket boats controlling the landings diverted the three waves of Ranger boats to the left because of the closing of Dog Green Beach. As A, B, and Headquarters Companies of the 2nd Rangers approached the shore, straddling the boundary between White and Green beaches, they too were hit by enormous volumes of fire directed at the LCA ramps.

Captain Dick Merrill, S-1 of the 2nd Rangers was first out of his Headquarters boat. Only he and one other of the first seven made it safely to shore on White Beach.[26]

Pummeled by breakers and enemy fire, mostly from WN 70, some of the grounded LCAs of Flotilla 701 broached and many of the 62 Rangers who were to survive out of the 160 or so who landed in Group Headquarters, 2nd Battalion Headquarters, A and B Companies, went over the side, using the LCAs, DD tanks, and other wreckage and obstacles as a shield against the enemy fire.

From his LCA, no more than one thousand yards behind the 2nd Rangers, Colonel Schneider observed what happened to the 2nd Rangers as they landed. Schneider was a veteran of Ranger landings in Africa, Sicily, and Italy. He knew that the 5th Rangers would never survive the holocaust that the 2nd Rangers had landed in. He asked his flotilla officer to shift the landing of the 5th Rangers about six to eight hundred yards to the east where he could see the beach was quieter. The British crews accomplished this maneuver despite being under increasing fire and despite being less than a thousand yards offshore. The 5th Rangers flotillas did not lose a boat and kept perfect formation.

Colonel Schneider ran a looser boat than Major Sullivan. As a result, Sergeant Graves and others were up and watching as the two waves approached the beach, broke out of the double columns, and slowly started to form a single line of boats. The interval between boats was about seventy-five feet. The first indication Graves had that they had been seen was when a large artillery shell landed in the water about one hundred feet to the side of the landing craft he was in. At that point, Colonel Schneider ordered all the men to stand up and told them to look straight ahead. "That is where you are going to land. Now get down and don't for any reason stand up."[27]

"The 5th Rangers' first wave of seven craft came in almost together, except for one of Company A's, slowed down by water trouble."[28] The touch down was around 0745, about ten minutes late. The Headquarters LCA landed at the extreme right of the line, about fifty yards from the boundary between Omaha Dog White and Red beaches. B Company's two boats landed to the left of Headquarters. E Company's two boats were next, and A Company's two boats were on the extreme left flank. The first wave was stretched over a front of perhaps 150 yards.

"The second wave was a little 'scrambled,' landing less than five minutes later"[29] *at 0750, just as three LCIs, 91, 92, and one other, landed. These LCIs were scheduled to land at 0740, the same time as the second wave of the 5th Rangers was scheduled.*[30] The second wave came in slightly to the right of the first wave, with the other half of Headquarters (my boat) on the extreme right. To my left was C Company. D Company was next, landing on top of Colonel Schneider's position. The second platoon of F Company overlaid B Company's position. The battalion front covered about two hundred yards overall.

chapter 6

THE BEACH

0745

The Beachmaster on Dog White had stopped his talk. We were about to find out why. We were still about a thousand yards out, and I had not seen the tragedy that struck A and B Companies of the 2nd Rangers and did not realize that we would land well to the east of our target beach. As we approached the beach, Sully ordered the men to keep their heads down, while he and, occasionally, I observed the developing situation. Schneider's wave landed to our left front.

> As his LCA touched down, Colonel Schneider was out of his boat in a flash, the rest followed as quickly as possible. The scene on the beach was almost unbelievable. Shells of all kinds were hitting at the water's edge. Machine gun and rifle fire were pouring in from the bluffs to the right. Rangers

were pouring out of the LCAs. Men were being hit as they left the craft, as they struggled across the beach. Dead and wounded lay in the water and across the sand. Those that could struggled to crawl out of the water before the advancing tide. On all sides, the wounded were screaming for medics. The noise was deafening. An LCM or LCT to the right was hit by artillery fire and burst into flames. Artillery was hitting all around while small arms crackled and popped. The top of the bluffs was obscured by smoke from a brush fire burning about halfway up.[31]

Schneider, followed closely by Graves, reached the seawall. The rest of Schneider's wave of LCAs touched down mostly to his left, and as the Rangers poured across the beach the force grew by the second. A Company of the 5th Rangers landed on the extreme left flank of the battalion. Even so, A Company found itself intermingled with other units, both Ranger and 29th Division, on the beach. Lieutenant Ace (Charles H.) Parker set about unscrambling the company and getting his squads and platoons back into some semblance of order. Reports showed A Company had had virtually no casualties crossing the beach.[32]

The coxswain of Parker's boat had put them off in waist deep water and then held his boat in and used his machine gun to try to suppress enemy fire and give his Rangers a better chance to cross the beach. Parker thought this an incredibly brave thing to do but wondered if it did any good since the Germans were firing from prepared positions located all over the bluffs. Parker located Colonel Schneider and reported to him, along with the two other company commanders in Schneider's wave, Captain George P. Whittington, B Company and Captain Ed Luther, E Company.[33]

For us in the second wave, still on the water and less than five minutes behind, the noise had now become deafening. An LCM or LCT to our right front was hit by artillery fire and burst into flames. Other artillery shells were detonating all around us, with the noise of small arms adding to the inferno. The scene was one from hell: smoke from the fires on the face of the bluff, fires from burning vessels and equipment, black ugly puffs from artillery bursting, dust and flying debris everywhere. A minute or so later, my LCA began to maneuver through the final obstacles. Another vessel, fifty to one hundred yards to our right, LCI 91, was hit by artillery. Waves lashed at us, throwing the boat right and left, pitching, tossing, smashing into the German obstacles. At one point we were crashing down on a pole-type obstacle with a Teller mine wired to it. Too bad, this is it! But another wave grabbed us, throwing us to the left, and we were past the mine and a few moments later, the rest of the obstacles.

> *Other boats had similar problems. One of D Company's boats disabled its ramp when it crashed into a pole. These Rangers had to climb over the sides of the boat to get off. Most craft grounded in water knee to shoulder deep. Only two were stopped by a sandbar so far out that the water in front of the ramp was dangerously deep (one man got off and was drowned). In both cases the Ranger officers had the coxswains drive in further.*[34]

Suddenly the coxswain gunned the engine and we hit the bottom with a jolt. {65949104} The ramp dropped and Sullivan dashed out to the left. I was second, chose the right, and shouted, "Headquarters! Over here!" The water wasn't as high as my boots. Our coxswain had done well by us. Ten yards of shallow water amid the damnedest racket in the world. You could hear the bullets screaming by. Somewhere an Oerlikon or a Bofors was beating out sixty rounds a minute at us. Rifle fire was coming from our right as was most of the machine gun fire. A DD tank to our right let fly a round. And now I was on the beach. It wasn't sand, more like gravel or small rocks, sloping upward. Forty or fifty feet ahead I came to a runnel of water. Machine gun bullets chewed the water up as I jumped into it. I remember thinking, "The Germans are trained like we are, bursts of three to five rounds, release and fire another burst. Jump in while the bullets are still splashing." I dashed into the splashes and yelled for the men behind me to keep moving. My runner, McCullough, was two behind me. I could feel him more than see him as he hesitated. The splashing stopped, he jumped ahead and was hit by the next burst. Then dry beach again. The seawall was ahead another fifty feet. It was packed with men two and three deep. They couldn't dig in because the rocks were six to eight inches in diameter layered deeply.

The seawall was made of logs, three or four feet high, with stone breakwaters running back toward the sea every seventy-five or so feet. Those breakwaters would prevent good lateral communication on the beach, though they did give us good protection from the flanking fire that poured down the beach from our right.

As I dropped down into the shelter provided by the seawall

and breakwater, I looked back. My men were coming up, dropping to the right and left of me. Artillery was falling at the water's edge, but only small arms from our right front was hitting near the seawall. Bodies were strewn all over the beach from the water's edge to the seawall. Other men were hiding behind the obstacles and wreckage near the water's edge. Still others were making the dash from boats to the seawall.

I tried to get my life belts off. They wouldn't come, jammed from something or other. I rolled over, still no luck. I could not go on like that so I got up on my knees and still no luck. I looked around. It was my first look at men in combat. They were huddled against the seawall as I had been only moments earlier, cringing at every bullet that cracked past. Artillery fire was churning the water's edge. To our left I saw LCI 92 touch down. As I watched, men started down the side ramps. Wham! An artillery round caught the starboard ramp. Must have hit a flamethrower there, for the whole side of the ship burst into flames that spread to the deck.

I looked back at our LCA. Men were still coming down the ramp. Because the ramp was so narrow, the men rushing out of an LCA had to go single file, and it takes a couple of minutes for all the men to clear the boat. Ah! There was Father Lacy, the last man coming out. He wasn't ten yards from the boat when . . . Wham! An artillery round hit the fantail of the LCA. Father Lacy was all right, but even though I couldn't see what happened to the crew, I knew. They had done their job well—too well, for the LCA was caught hard on the beach, too hard to back off. I looked away. My first dry landing, well, almost dry, and the British crew paid for it with their lives.

I didn't see Father Lacy again on D-Day, but others saw him and sang his praises. Lacy didn't cross the beach like we heroes did. He stayed down there in the water's edge where the artillery was hitting every boat in sight, pulling the wounded forward ahead of the advancing tide. He comforted the dying. He calmly said prayers for the dead. He led terrified soldiers to relative safety behind debris and wreckage, half carrying them, half dragging them, binding up their wounds. Never once did he think of his own safety, always helping those that needed his help to survive that awful inferno.

When the 5th Rangers left the beach a few minutes later, Father Lacy stayed behind at the water's edge, doing the work for which God had chosen him. True to his word the Padre caught up with us later. He was delayed, he said. (See Appendixes B and D for more on Father Lacy.)

By now my men were dropping around me and into the adjacent bays. I yelled to a radio man, I think it was Corporal Soper, who stood up and cut my preservers off.

"Anybody hit?" I shouted. "Yea, McCullough got a slug in the back of his leg." One man, my messenger, only two behind me was hit. Not bad for thirty-three men crossing a hot beach like that.

The other companies had much the same luck. With about 450 Rangers crossing the beach, there were only four or five casualties. C Company, in the second wave, did have problems with the 81-mm mortars and was slow getting the equipment ashore; having to float in some of the mortar ammunition on life preservers.[35]

I called for Sullivan, "Over here, Red," was his reply. He was in the next bay to my left (east). I slipped around the breakwater and made my report, one casualty with the rest of Headquarters dispersed in these three bays. "Sully, for God's sake do something; we can't sit behind this seawall, taking all this fire for long." We immediately passed the word for Schneider. He turned up about fifty yards to our left giving orders to the company commanders. Sully told me to take charge of this flank, while he went over to Schneider to report and find out what was going on.

I began checking the men, making sure they still had their weapons and ammunition, and getting them more or less collected for the next move, all the while wondering what that move was to be. The part of Headquarters Company that had landed in my boat consisted of most of the Communications platoon, some medics and a few drivers and cooks brought along as runners and messengers. The battalion staff sections and the Joint Assault Signal Company's detachment made up the bulk of Colonel Schneider's boat. The rest of Headquarters Company would land about midnight along with our vehicles, bedding rolls, and equipment.

As I continued to look around me I observed several small parties of infantry had left the beach to my right. Apparently, some infantrymen of the 116th or Rangers from the 2nd Battalion had worked their way off the beach and were about halfway up the bluffs to our right, engaged in a firefight.

Company C, 116th Infantry landed at 0710 on Dog White Beach, ten minutes early and nearly a thousand yards east of the Vierville exit where it was supposed to land. This mis-

landing was a godsend, for the beach was nearly clear of enemy fire, thanks to the heavy smoke from brush fires, and C Company being the first unit to land on that sector of the beach. C Company lay low behind the seawall reorganizing, taking care of its wounded and injured, and trying to get its bearings while waiting for something to happen. One of the injured was the company commander, Captain Berthier B. Hawks, whose foot was crushed by the ramp of his landing craft. Despite his severe injury, Hawks would continue to lead his company for the next several days.

At 0750 Private Ingram E. Lambert crossed the beach road with his bangalore. It failed to fire and Lambert was killed, but his action ignited the company. Lieutenant Stanley M. Swartz jumped up to take Lambert's place, got the bangalore to detonate, and C Company, 116th Infantry burst through the gap. About ten minutes later, C Company had reorganized in the flat beyond the wire and was on the attack up the bluffs. Another gap was blown and now two groups were on the way uphill, probably with some of the 2nd Rangers with them. They carried the top of the bluffs to find the Germans had abandoned the trenches there. The company was stopped by enemy fire after a two-hundred-yard advance into some flat open fields.[36] Lieutenant Swartz was killed in this action later. This was the attack that I observed to my right front shortly after landing. Since General Cota was in that vicinity when the attack began, I'll always feel that Lambert and Swartz acted under Cota's urging.

For the first time I began to take in the terrain in front of us rather than what was happening on the beach and at the seawall. The terrain was different from the maps and models. The high steep hill was 100 to 150 yards in front of us, not right down on the beach where we thought it would be. The top half was covered with smoke and flame from a brush fire that had started to our right. The terrain was flat from the foot of the hill to the beach road just in front of us. *(Some forty-five years later, Fran Coughlin, Headquarters Company, 5th Rangers, told me he reached over and touched the road. It was blacktop.* Omaha Beachhead[37] *confirms that fact.)* Next was a battered little stone wall and then the wooden seawall. I looked at the seawall again. My God! This was the wrong beach! Our beach had a stone seawall, not a log seawall. Where were we? We couldn't be to the right of Vierville because there would be cliffs in front of us and the Pointe et Raz de la Percée beyond that. Therefore we must be to the left.

The next seawall going east from Vierville, according to the maps and sand tables we studied so hard, started near the eastern edge of Omaha Dog White, a wooden one. I looked around even more carefully, the seawall and the breakwaters ended three or four bays to my right. Farther down to my right I could see one, perhaps two, DD tanks of the 743rd backing down to the water and then slowly coming across the beach, each time giving five or six men cover to cross the beach. Back and forth, but that was two to three hundred yards away. As a guess, we had been on the beach at this point no more than ten minutes, and probably less. Clearly, we were on or near the boundary between Dog Red and Dog White beaches.

Not ten yards to my right, a grizzled old Engineer sergeant set a heavy machine gun tripod down in a gap in the stone breakwater. He then ambled back to my left. A moment later he returned, cradling a heavy water-cooled machine gun in his arms. A thin Engineer lieutenant in a green sweater was carrying boxes of ammunition and cooling water for him. Together they very calmly set up their machine gun in that exposed gap in the breakwater. The sergeant very methodically began to traverse and search the hill to our right where the firefight appeared to be. The lieutenant, and I'll always remember the disdain he showed, turned around, standing on the breakwater with his hands on his hips and spat out something to the effect, "and you call yourselves soldiers." He then tried to organize his men and then the stragglers from the 116th Infantry, but to no avail. *Omaha Beachhead*[38] places this event, or at least one remarkably similar to it, on Easy Red Beach.

Now that the second wave had landed, the rest of the commanders—Captain Bill Wise, C Company; Lieutenant George Miller, D Company; Lieutenant John Rahmlow, 2nd Platoon F Company; and Major Dick Sullivan—were arriving at Colonel Schneider's position to give their reports. After assessing the situation, Schneider gave the "Tallyho," a code word for orders to proceed to the rendezvous points near Vierville by platoon infiltration.

Before the companies could move off the beach, "certain preparatory moves had to be made. The mortars were ordered to set up back of the seawall and get ready to fire on the bluff."[39] Bangalore torpedoes had to be assembled and placed under the wire on the far side of the beach road. Orders had to be given on the order of units going through the gaps in the wire, formations after clearing the wire, and objectives and missions.

Lieutenant Ace Parker, having the farthest to go, immediately moved back to A Company and gave Hathaway, his bangalore torpedoman, orders to blow a hole in the barbed wire that barred passage to the high ground beyond the beach. After the bangalore had done its work, Parker, followed by A Company, dashed through the gap, across the flat, and began to climb the bluffs to his front. The smoke from the brush fire had two benefits, it obscured A Company's advance from the Germans dug in at the crest and it burned away the brush, exposing many of the mines emplaced on the flat area in front of the bluffs and on the slopes of the bluff itself. The smoke also served to disorganize the movement. Many Rangers had to put on their gas masks to get through the smoke and some were momentarily lost. Contact between platoons and sections was rarely maintained. Lieutenant Suchier, A Company, went into the smoke leading his platoon and came out with only two men.[40]

Sullivan worked his way back to my vicinity and passed the word to me that Schneider had given the "Tallyho." The gap in the wire that I was to use was in front of D Company, three or four bays or seventy-five yards to our left. Headquarters was to follow behind C Company's machine gun sections. As I got ready to move out, I saw, or had pointed out to me, a man casually wandering down the beach toward us.

It was clear that he was someone with authority, for he was shouting orders and encouragement to the troops huddled against the seawall or burrowing into the edge of the embankment. By the time he got to our vicinity, I realized he was quite high ranking, a colonel or general. I jumped up, ran over to him, saluted, and reported, "Captain Raaen, 5th Ranger Infantry Battalion, Sir!" "Raaen, Raaen, yes, you must be Jack Raaen's son. I'm General Cota. What's the situation here?" "Sir, the 5th Rangers have landed intact here and to the east about two hundred yards. The battalion commander has ordered us to proceed by platoon infiltration to our rendezvous points." He asked where Colonel Schneider was located. I pointed out the bay where Schneider was sitting against the seawall, still talking with one or two of the company commanders. I offered to take him to Schneider, but he said, "No. You stay here with your men." He started toward Schneider, then stopped and turned to the troops in my vicinity and said, "You men are Rangers. I know you won't let me down." And with that, he was off to see Schneider.

Before Cota reached Schneider, a flurry of artillery fire caused him to hit the dirt. Tom Herring, C Company, 5th Rangers, was lying to the left of William Stump, also C Company. Stump asked Herring for a match, saying his were wet. "Mine too," said Herring. Stump reached across Herring's back and punched a soldier next to Herring and asked, "Hey, Buddy, you got a light?" As the soldier rolled onto his left side, the star on his jacket epaulet was visible to both Stump and Herring. Stump said, "Sorry, Sir!" Cota reached into his jacket, pulled out a Zippo, flicked it, held it for Stump to light up and said, "That's OK, son, we're all here for the same reason." Shortly after that, Cota rose and began pacing to and fro on the beach, repeatedly yelling, "You have to get off the beach or you're gonna die." Shortly after that, he reached Schneider's position.[41]

When Cota reached Schneider, Sergeant Jim Graves observed that Schneider jumped up at attention and saluted the general. Cota meticulously returned the salute as he had mine. Graves thought it foolish for anyone to expose himself so unnecessarily to that dreadful small-arms fire coming in from our right.[42]

Cota asked, "Are you Colonel Schneider of the Rangers?" "Yes sir!" Schneider replied. Cota then told him, "Colonel, you are going to have to lead the way. We are bogged down. We've got to get these men off this God-damned beach." Graves couldn't hear the rest of the conversation, but it was clear that Cota was changing the mission of the 5th Rangers. The battalion was now going to lead the way off the beach and that way was an attack straight up the bluffs.[43] *Lieutenant Shea, General Cota's aide-de-camp, reported that Cota ordered Schneider to blow the wire and lead the 5th Rangers against the enemy fortifications at Pointe et Raz de la Percée. It was these forts, lying between the Vierville exit and the Pointe et Raz that were shelling the assault boats as they landed on the Omaha beaches.*[44]

chapter 7

ASSAULT UP THE BLUFFS

B arring the way to the attack up the bluffs was a broad band of barbed wire that seemed to run the whole length of the beach road. The double apron wire was emplaced just beyond the seawall and on the far side of the blacktop beach road. The wire was impenetrable, and the only way to attack through it was to blow holes in it. That meant bangalore torpedoes had to be assembled and gotten under the barbed wire.

Bill Reed, D Company, 5th Rangers, and Woody Dorman, though a team, had separate torpedoes. Each assembled his five-foot sections, using the seawall and breakwater for protection from the fire coming in from the right. As each section was attached to the next with a cylindrical sleeve, the assembly was pushed out onto the road. A pull-type fuse lighter with a four- to five-second delay was attached to a blasting cap and put into the base of the last section of the torpedo. Trouble was, the barbed wire was on the far side of the blacktop road and only one section of the torpedo reached the edge of the wire. To solve this problem, both men jumped up, running across the road, pushing their bangalores ahead of them. When the torpedo was under the whole stretch of barbed wire, they pulled their fuse lighter toggles and ran back across the road, jumping down to safety behind the seawall. Both of D Company's torpedoes went off almost simultaneously, blowing a double-wide gap.[45]

After General Cota finished speaking with Colonel Schneider, he turned toward the men nearby and said, "Rangers! Lead the Way!"

"After receiving the 'Tallyho' order from Colonel Schneider, Captain Ed Luther of E Company had moved back to his company and was giving orders to his two platoon leaders, Lieutenants Dee Anderson and Woodford Moore. An officer walked along behind his men and started urging them to get across the wall. Looking over his shoulder, Luther put up a warning hand and said, 'Hey, Bud! Take it easy, don't get excited. This is my outfit and I'll take care of it!' The officer called out, 'Well, you've got to get over that wall!' Luther hollered back, 'Quit bothering my men; you'll disorganize them. The Colonel's over there if you want to see him, but quit bothering me!' Just then E Company's bangalore exploded and the platoon leaders started to cross the wall. A big grin came over the officer's face, and he started walking down the beach again. As he turned, Captain Luther saw the star on his shoulder for the first time, and took off over the wall hoping that General Cota wouldn't remember him."[46]

"The movement off the beach under the 'Tallyho' order was essentially by platoons. E Company started off with its two platoons together in column. Most platoons used a formation of sections in column and line of skirmishers within the section."[47]

Within moments after Woody Dorman and Bill Reed, D Company, had blown their bangalore torpedoes, Lieutenant Francis Dawson, D Company, dashed through the gap, leading his platoon in a charge across the flat and up the hill. (This is the gap in the wire that D Company, C Company, the Command Group, and Headquarters Company would pass through.) Dawson was halfway up the hill when he looked back and found out that he was alone. His platoon had been slowed by their struggle to get up the wall and through the gap in the wire in single file. After the platoon caught up, they were slowed again near the top of the hill when they had to clean out a few Germans from a trench system along the bluff edge. Reed, slightly behind Dawson and to his right, fired a rifle grenade that hit right in the pit of a machine gun position. Dud! Dawson immediately jumped up and charged the position, firing his sub-machine gun and silencing the machine gun that had been firing along the beach.[48]

At that point, B Company under Captain George P. Whittington Jr., which had been following D Company closely, passed through Dawson's platoon and carried the crest.

Upon reaching the top of the hill, Captain Whittington immediately ordered Sergeant Mac McIlwain (Walter N.) to take a patrol and try to locate the commanding officer of the 116th Infantry. The battalion needed this information badly, since General Cota had ordered the 5th Ranger Battalion to assist the 116th Infantry in establishing the beachhead. During the attack up the bluffs, Colonel Schneider had lost contact with General Cota and now needed to establish contact with his immediate commander, Colonel Canham, commanding officer of the 116th Infantry.[49] Now that the troops were beginning to move off the beach, General Cota followed them through the D Company gap in the wire and up the hill.

After a considerable time working to the right in the fields between the crest and the first hedgerows, McIlwain's patrol made contact with Colonel Canham and a handful of officers and enlisted men near Hamel au Prêtre. Colonel Canham indicated he had "had no contact with his units and did not know where they were located." Canham asked that the patrol stay with him to help provide his small command group with better security. This was the first time that a small unit of B Company, 5th Rangers, would be asked to do this. Later, in Vierville, Canham would make a similar request of another B Company unit. McIlwain declined to stay, since his orders required him to report Canham's position back to Whittington as soon as possible.

On the way back to the company, about 0930, the patrol was hit by heavy artillery fire. McIlwain was wounded, T-5 Elmo E. Banning was killed, and Bernard Akers was wounded. This was probably the same fire that caught me and Headquarters in an open field at about the same time.

The patrol made it back to the company, which, by now, had advanced to the coastal road and was about to launch an attack across the road toward Vierville. Because Colonel Schneider was at the head of the column with Captain Whittington at the time, the patrol was able to make its report to the two of them.[50]

"Fortunately, enemy opposition along the face of the bluffs was very light. There were no occupied enemy positions of any consequence along the 5th Rangers advance up the bluffs. The advance of C Company of the 116th Infantry and of the 2nd Rangers and the blinding smoke cut off any German observation of the advance. The bluff slope, while steep, had many small folds and irregularities, which gave cover against occasional fire from the right flank. There was no trouble with mines. All told, the 5th Ranger units lost only eight men from the wall to the top of the bluff."[51]

"The only German positions reported were on the left (east) flank. Here, D Company's second platoon under Lieutenant Bob Stowe found a trench system running along the bluff edge. The platoon moved east through the trenches, cleaning out an occasional sniper, and accounted for eight enemy. They found one machine gun position firing down on the beach and drove the Germans out with a rifle grenade, thereby helping some Ranger units that were coming up the bluffs just below."[52]

"The move up the bluffs took about fifteen minutes. By 0830, most of the 5th Ranger units were up or arriving.

*Lieutenant Suchier, on the extreme left flank, found two
116th men out in a field. Lieutenant John Gawlor, C
Company, found Captain Hawks, the commander of C
Company, 116th Infantry, at the top of the hill. Hawk's foot
had been crushed by the ramp of his LCA."*[53] *Nevertheless,
Hawks got up to the top with his men and was waiting for a
chance to move them inland.*

Meanwhile, responding to the original orders to infiltrate,
I and my half of Headquarters shifted to the left, leaving the
Engineer lieutenant with his hands still on his hips looking
disgusted. In the process I picked up the half of Headquarters that
had landed in Schneider's boat. Among them were my second in
command, Lieutenant Howard Van Riper, and Sergeant Graves.
Graves, like all the rest of us, was confused by having landed on
the wrong beach. He asked me, "Captain, do you know the way
to the rendezvous point?" "Yes, Sergeant, I believe I do." Graves
looked relieved as he said, "Then I will try to follow you."

We found the gap. A line company was passing through. Some
Heine was firing from the right along the beach road. There was a
shattered stone building, probably a pill box, just across the road.
C Company was moving through now. I tagged on, rushing across
the road. There, lying stomach down and trousers down on a stone
slab on the left of a ruined pill box was little Tony Vulle. Tony,
though the smallest man in the battalion, insisted on carrying the
heaviest load, the 81-mm mortar baseplate. Vulle was now having
general repairs done to his gluteus maximus while the battalion
moved by. He hadn't crossed the road fast enough.

We trotted down a little path and then as the column stopped, hit the dirt. It wasn't too comfortable there in the open, so I shifted my men to the left into a small gully or ditch. The column moved again, stopped, then moved again. There was heavy brush at the base of the hill and a flagstone path leading through it. Starting up the hill, there were about six stone steps straight up and then a path leading up and right. The column stopped as I reached the last step. We were well out of any small-arms fire, screened by the hill itself and the smoke from burning brush on the slope. I sat down and looked back toward the beach. Men were still coming through the gap in the wire, probably from the 116th. Boats were still coming into the beach. As far as my eye could see, our gap was the only one through the wire.

Actually, the 5th Rangers blew four gaps in the wire. Hathaway of A Company blew one. Dorman and Reed, D Company, blew another. Still another was blown by Beccue, B Company. The fourth gap was blown by E Company. A fifth hole and a sixth were blown by C Company, 116th Infantry, as a prelude to their attack up the bluffs. (I saw this attack shortly after I had reached to the seawall, and it was this attack that the Engineer sergeant was assisting with his machine gun fire.)

Somewhere on the way up the steep bluff, I learned that the battalion's orders had been changed from platoon infiltration to attack as a battalion.

"When Colonel Schneider reached the top of the bluffs, considerable confusion prevailed. The first Ranger units had come up into the fields more or less disorganized as platoon leaders chose their own routes to their rallying points. Colonel Schneider decided to pause in the advance,

reorganize, and then proceed as a battalion.

"Runners were sent out to locate units, officers came in and reported. B Company and C Company seemed to be the only companies with all their units accounted for, at the moment, and B Company was ordered to lead off, finding its own route to the assembly point. Vierville itself was to be avoided. The other units were to follow in column of companies, with platoons abreast and sections in column.

"The remnants of Headquarters and A and B Companies of the 2nd Rangers were coming in from the west and were fitted into the column as a provisional Ranger Company under the command of Captain Edgar L. Arnold, the B Company commander. The 2nd Ranger companies were too reduced in numbers to serve as advanced and flank guards, their original mission, and were placed in 5th Battalion reserve. C Company set up its 81-mm mortars along a hedgerow to support the advance. At least half an hour was taken up in making these arrangements." (I was not aware of any such reorganization. To me, the delays were just the way things went when you fought in column on a narrow front.) "There were no signs of enemy south or west. A small group to the east had been fired on as they pulled out of a position beyond the first hedgerow."[54]

The advance had slowed to a crawl. Climb a few steps, stop, and wait. Climb some more, stop. The reason was understandable. Most of the fields beyond the crest were either heavily mined or had signs indicating they were mined. Advancing on a broad front invited many casualties from these mines. Moving in a column reduced the chances of casualties from mines considerably, but in the process, progress was slowed.

> *The change in orders caused at least one problem. When "Tallyho" was ordered, A Company, led by Ace Parker (Lieutenant Charles), blew a hole in the wire and took off up the hill. Parker and his men reached the top of the hill essentially unscathed, but as they continued the six hundred yards through the hedgerows to the blacktop road, Lieutenant Oscar A. Suchier Jr., leading the second platoon, was hit by rifle fire. Unbeknownst to Parker, that held up the rear half of his column while Suchier's arm wound was tended to. Parker and his twenty-three men kept going toward his rallying point, the Chateau de Vaumicel, {64629062} just south of Vierville. The rest of A Company, those behind Suchier, learned of the new orders to stay in battalion formation and did not follow Parker.[55]*

We now had lost two company commanders and two platoons, the one from F Company in the LCA that swamped and now one from A Company. (Actually, Parker's rallying point was the Ormel Farm. The mapmakers had made an error and placed the name Vaumicel on the wrong location. The Chateau de Vaumicel, a classical stone chateau, is about five hundred meters to the west of the farm on the Englesqueville Road.)[56]

The column ahead resumed its move up the steep slope and into the smoke from the burning brush. Our path continued diagonally to the right uphill, but after about fifty yards we had to sidestep up the steep slope. By now the smoke was so bad that we found ourselves gasping for breath, but gulping in smoke. I couldn't see ahead through the tears. Some of the men nearby had been asking if they could put on their gas masks, but I had decided to tough it out. Now even I gave in and passed the word for gas masks. We had the new assault masks with the canister on the left side of the face piece. My mask wouldn't come out of the carrier, being jammed somehow. I put my helmet between my knees and then yanked hard on the mask. The mask came out, but so did the maps and the D bars I had secured in the carrier. Fortunately, I was able to retrieve everything but the orange that disappeared downhill.

I finally got the mask on, put on my helmet, and took a deep breath and almost smothered. In recovering my maps and D bars, I had forgotten to take the covering plug out of the canister. I yanked off the mask, and with that my helmet came off and started to roll down the steep hill. Sergeant Graves grabbed it. Now I was choking in the smoke. I finally got the mask and helmet back on, yanked the tab, breathed in some dank smokeless air, took three steps, and was out of the smoke. I was so furious, I kept the mask on another fifty feet just to spite myself.

We left the path now (it angled back to the left past a little shack) and continued to the right to the top of the hill. We saw our first German, a dead one. He was lying in a little hollow just below the crest. None of us had ever seen a dead man before, at least this close. He was sort of greenish yellow, looked like wax. Of course, he was a booby trapped dummy! It wasn't till much later that we realized our wax man really was a dead man. In the hollow, we paused for breath before crossing a tiny stone wall into the hedgerow country.

Just before stepping over the wall I looked back and took in the scene below once more. Nothing had changed. Boats were still landing troops. Artillery was still hitting near the water's edge. Small-arms fire was pouring in from the west flank, and infantry troops were struggling through the gaps in the wire to join in the climb up the bluffs.

chapter 8

THE HEDGEROWS

0830

There were two connecting files from C Company at the wall. They gave me instructions on the battalion route and how the company should proceed to avoid the mines and the most dangerous enemy fire. I replaced the connecting files with two of my own men. After crossing the wall, we moved to the right, parallel to the beach perhaps fifty to a hundred yards from the edge of the bluffs. Along the near edge of a north-south hedgerow I found C Company's 81-mm mortars and a light machine gun section emplaced and prepared to fire parallel to the beach. I dispersed Headquarters in the field behind the mortars and left a non-com in charge while I went to find out what was happening. Just as I left, Van Riper came up with the rest of Headquarters and dispersed into the same field.

There was some low-velocity German artillery passing close

overhead, heading for the beach. A few hundred feet inland, to the west and south of us there was intense rifle and machine gun fire as the advancing Rangers drove the Germans back, hedgerow by hedgerow. Occasional grenades, mortar rounds, and artillery added to the noise. Only scattered small-arms fire reached us on the crest itself.

Captain Bill (Wilmer K., commanding officer, C Company) Wise told me I would find Sullivan and Schneider at the southern edge of the hedgerow but not to go into the open field beyond, because the enemy was to our front. I found Sergeant Graves and his radio along the edge of the hedgerow near the gate at the end of the field. I plunked down beside him to ask where Schneider was located. Graves asked me what I thought of the situation. According to Graves, I answered, "Kind of exciting isn't it." Graves then pointed out Schneider and Sullivan just on the other side of the gate at the end of the field. {vic 65789090}

Schneider was occupied so I spoke with Sullivan. Unfortunately, there was no known situation for Sullivan to give me. All he could say was because of the minefields the battalion was moving in a column formation and that he had seen a patrol move off to the southwest along the fence toward the far hedgerow. Sullivan had me move out along that fence to see if I would draw fire, simply because that was the best route to move the portion of the battalion that had not yet displaced along the crest. I zig zagged about seventy-five to one hundred yards before I reached cover. {65559081} I had drawn enough fire to mention, most of it, friendly. There was a dead German in the hedgerow.

A COMPANY'S ADVANCE

*A*t the blacktop road, about one thousand yards from the bluffs, Ace Parker and his twenty-three men from A Company continued west toward his rendezvous point at the Chateau de Vaumicel (actually the Ormel Farm), just south and east of Vierville. A small party from E Company, led by Lieutenant Woodford O. Moore was with them. The force crossed the St. Laurent-Vierville Road (the coastal blacktop road) and were within a few hundred yards of the farm before meeting any enemy resistance. As they started to cross an open field they were taken under fire by a group of enemy lodged in some trees at the far end of the field. All hit the dirt, except T-5 William J. Fox, Parker's runner and sniper. Instead, Fox crouched down and was hit in the shoulder. Lieutenant Moore was also hit in the head. Parker was carrying a musette bag, loaded with everything he thought he might need for the next week. Even though Parker was protected by a crease in the ground, his musette bag stuck up high enough that it was hit several times by rifle fire. He managed to get the pack off, and he and his men crawled into a north-south ditch where they were protected from the rifle fire.

They roped and then dragged their wounded into the ditch as well, tended to their wounds as best they could, and, leaving them with extra water, began to work their way down the ditch toward the chateau. Moore survived, Fox died of his wounds. The platoon reached the chateau (Ormel farm) {64629062} around noon, after spending about three and a half hours in the ditch, working its way past this resistance.[57]

B COMPANY, 116TH INFANTRY LANDS
(First Wave)

Two boats of B Company of the 116th Infantry had been hard hit as they landed at the Vierville exit right on top of A Company/116 at H+30. The coxswains of the other four boats, seeing the disaster that befell A Company, drew back and veered right and left away from the A Company shambles. One of these carrying an infantry boat section commanded by Lieutenant Leo A. Pingenot landed near C Company of the 2nd Rangers in its attacks on the enemy positions at Pointe et Raz de la Percée. A Ranger, stationed at the ropes running down the cliffs to the beach, saw them and an officer of C Company sent a runner to the platoon, which immediately joined C Company in its attacks on the trenches and bunkers lying in front of WN-73 and the more distant Pointe et Raz de la Percée.

Another LCA of B Company of the 116th landed just to the right of the Vierville exit. This boat section, led by Staff Sergeant Robert M. Campbell, was not so lucky. It was cut to pieces by enemy small-arms fire.

In Lieutenant William B. Williams's boat, the coxswain steered sharply left and moved along the coast about a mile, then put the boat straight in. The ramp dropped on dry sand and the boat section jumped ashore. Mortar fire had dogged them all the way along the coast, and as they cleared the ramp a mortar round scored a direct hit on the LCA. But dry landing or no, the men were so weak from seasickness and fear that they literally crawled across the beach, dragging their equipment behind them. After twenty minutes, Williams and ten men made it to the seawall. Five others had been wounded and six were missing.

This boat section managed to fight its way off the beach and attack a fortified position near Les Moulins. Lieutenant Williams was wounded in that fight and ordered Technical Sergeant William Pearce to lead the section, but to "go the other way toward Vierville." After another fierce fight, the section reached Vierville a little after 1000.

The LCA carrying another B Company boat section swung left and landed on Dog White Beach just below Hamel au Pretre. Lieutenant Walter Taylor led his section across the beach, suffering six casualties. Then up the bluff and into Vierville where in a two-hour fight the section whipped a German platoon without losing a man. At that point, Taylor was joined by Sergeant Pearce, and under Taylor, the platoon, out of contact with other units, turned south toward the Ormel Farm. Taking fourteen prisoners along the way, it met heavier resistance south of the chateau, and because it had no machine guns or mortars was forced to withdraw to the farm and managed to hold the grounds with rifle fire. Here they were joined by Parker and the platoon from A Company, 5th Rangers, about 1200. Neither unit knew there were any other units near Vierville. A short time later, the platoon from Company B, 116th Infantry withdrew to Vierville in search of other elements of the 116th. Parker and his 1st Platoon remained at the farm waiting for other Ranger units to catch up.[58]

5TH RANGER ADVANCE TO THE COASTAL HIGHWAY[59]

B *Company of the 5th Rangers was leading the battalion after it cleared the crest. Being the first unit there, it did not suffer as much from intermingling as the other companies. Colonel Schneider ordered B Company to lead the attack toward the coastal road. By now it was between 0900 and 0930. A hedgerow ran south from the edge of the bluffs where the Rangers came up the hill. The column being formed was grouped along this hedgerow when B Company's 1st Platoon started off as point with Lieutenant Bernard M. Pepper in command.*

With scouts out and flank protection, Pepper moved his platoon generally west toward Vierville. However, after a short distance, naval gunfire began landing directly in their path. Lieutenant Pepper wisely turned inland, more toward the south, and reached the blacktop coastal highway about 0930.

The platoon's point was fired on by snipers and machine guns from across the highway. Some scouts were sent out across the road to locate the machine guns.

The scouts located one machine gun in the southwest corner of the large wheat field in front of them. Colonel Schneider, who was at the head of the column with Captain Whittington, ordered Pepper to take out the gun. As Pepper and Whittington were crossing into the wheat field with the leading section, they saw about twenty Germans leaving the area where the machine gun had been located. Thinking the gun position had been abandoned, they started across the field only to be fired on by the machine gun. Fortunately the wheat was about two feet high and provided fairly good cover and the platoon was able to traverse the field and attack the position from both sides of the hedgerow leading to it. Two Rangers who had crossed the hedge found the rear of the German position and killed all seven enemy in the position.

A few moments later, German mortar fire began to land along the hedgerow they had used and Captain Whittington ordered a hasty withdrawal to the highway. They made it to the road just one step ahead of the mortar fire.

Back on the highway, another machine gun somewhere near the southeast corner of the wheat field opened up on B Company. Apparently, this gun had not been able to fire to the west and could not interfere with Pepper's attack through the wheat field. But now, firing on the road, it knocked out the platoon's BAR (Browning Automatic Rifle) team.

Thinking the battalion would follow, Captain Whittington decided to abandon his efforts to envelop Vierville and attack directly toward the village along the coastal highway. For a time, a platoon-sized unit, probably from C Company, 116th Infantry, followed B Company for about a half mile. It then dropped behind.

Moving straight down the highway without opposition, B Company's point reached Vierville by 1100 hours, apparently just after the platoon from B Company 116th had passed. Not a shot was fired in the village, and there were no signs of enemy or friendly forces.

Captain Whittington, still pursuing the 5th Rangers

mission, pushed through the village toward Pointe du Hoc. By noon, B Company's point was about five hundred yards beyond the village. There it ran into prepared enemy positions. The enemy waited until the point was past their positions and then opened fire on Pepper's main body.

Pepper deployed his platoon in a field south of the road to attack a machine gun position and some snipers there. Suddenly, up the road from Vierville came General Cota, all alone and smoking his inevitable cigar. He asked Pepper what was holding him up.

"Snipers," said Pepper.

"Snipers? There aren't any snipers here." At that moment a shot came close to the general. "Well, maybe there are," said Cota and he walked away.

B Company was now fully engaged with fire from both flanks and the front. Whittington pulled the company out of that untenable position to some houses where the 1st Platoon went into defensive positions and the 2nd Platoon went off to the south to try to envelop the enemy positions.

Lieutenant Gregory took the 2nd Platoon of B Company south along a hedgerow where a heavy weapons unit of the 116th was in position. Two rounds of naval gun fire inflicted two casualties and caused a short halt. Resuming its flanking movement, the platoon went south to the first dirt road and there met Colonel Canham, the regimental commander of the 116th Infantry. This time, Colonel Canham prevailed and the Rangers were promptly impressed as protection for Canham's Command Post (CP). During the rest of the day, the platoon was involved in several minor firefights.

After I made my way back to Sullivan, I rejoined the company as the column began to move out, using the fenceline I had just reconnoitered. As we got to the hedgerow, we moved to our right until there was a break. Then through the break, across a field, and to the cover of another hedgerow. We continued moving this way, in bounds along hedgerows and down little paths between hedgerows, until we came to a large irregular field. {vic 65409070} There seemed to be no small-arms fire so we dropped to the ground to rest and reorganize. By now, the time was between 0900 and 0930.

Suddenly, I heard a low whine, not at all like the noise the artillery had been making as it passed over our heads toward the beach. No, this noise was coming straight at us. We all hit the dirt as four or five shells detonated in our field about thirty yards away. A quick check showed no one was hit. There was another unit in the field beyond the impact, but they seemed to have weathered the incoming as well. Another incoming salvo meant this field was unhealthy, but there was no place to go. By the third or fourth salvo, we had reached the point where we could estimate where the shells were going to impact and with that our confidence went back up. Half the time we didn't even duck. Since this shelling occurred about 0930, it was probably the same artillery fire that caught McIlwain's patrol as it returned to B Company after establishing contact with Colonel Canham. Glassman's *Lead the Way, Rangers* covers this artillery fire by saying that there were several direct artillery hits on the rear of the column, causing many casualties.[60]

chapter 9

BATTLE ALONG THE COASTAL HIGHWAY

---

*D*uring the time that the rear of the battalion was being shelled, the forward elements of the main body
had reached the coastal road about a mile east of Vierville.
{vic 65309050} Before the invasion, Colonel Schneider had
given Sergeant Jim Graves a list of the call signs of other
units in the invasion together with their radio frequencies. When the battalion cut the coastal road about 0930
or 1000, Schneider had Graves transmit a radio message
to General Bradley aboard the USS Augusta saying, "Have
reached the top of the hill. Have cut the blacktop road. Will
attack towards Vierville-sur-Mer." If Bradley saw that message, it was probably the first good news he had received
from Omaha Dog Beach.

    Force C had gradually been losing units. One platoon of

Company F had swamped coming in to the beach and was landed elsewhere. The 1st Platoon of A Company and most of a platoon from E Company had moved inland, not aware that the "Tallyho" order had been revoked. B Company had attacked down the blacktop road toward Vierville, thinking the rest of the battalion was behind it, and then "lost" a platoon to guard the CP of the 116th Infantry Regiment.

But the main body did not follow B Company. Other 5th Ranger units had joined B Company in its attack across the coastal highway, and when B Company disengaged and started west, the rest of the battalion was becoming embroiled to the east.

While B Company was attacking the machine gun positions just across the blacktop road, E Company was ordered to make a flanking attack to the east in support of B Company. Captain Luther sent a patrol of one section south from the blacktop, one hedgerow east of the wheat field. The section was fired on by a machine gun in the southeast corner of the field.

The patrol ducked into the hedgerow ditch and worked south, until it met fire from a machine gun ahead of it near the same hedgerow. Lieutenant Dee Anderson was killed in this firefight. His patrol got near enough to use hand grenades but couldn't land them in the German holes. Meantime, Captain Luther went with another section east to the next hedgerow to outflank the newly located German machine gun positions.[61]

The column in front of me began to move out again. Firing

ahead showed that the attack had begun and that it was meeting resistance. With headquarters out of that field and in the comparative safety of a path between hedgerows, I went forward again. By now it was well past noon. I don't remember whom I met, but one of the lieutenants of the 5th gave me the situation in a nutshell.

Schneider had ordered an attack across the blacktop road that would sweep to the southwest toward the Chateau de Vaumicel (actually the Ormel Farm). The attack by B Company was held up by intense machine gun fire from positions in a hedgerow to the left front.

That's easy. Schneider sent a platoon to attack the machine gun from the east flank. This attack was held up by another machine gun position still farther east. Another machine gun and another envelopment, and now the 5th Rangers had four companies spread out along the road to the east. This was the wrong direction! Clearly Schneider had not expected the interlocking defenses that the enemy had set up south of the road.

These hedgerows were like nothing any of us had ever seen before. We had expected hedges such as we saw in England or even at home, but no way! A Norman hedgerow was a long thick mound of earth, six to ten feet high. On top of the mound was usually a hedge of some sort, old, gnarled, heavy, with huge roots holding the mounds together. We later found out that tanks could not drive through these monsters. The Germans dug holes in the back side of these natural barriers and hollowed out small machine gun nests and fighting positions. Because of the natural camouflage and the overhead cover, it was impossible to see these positions as they fired at you. The fields between the hedgerows were usually rectangular and of all sizes from twenty or thirty yards across to perhaps one hundred acres, with the average about ten acres. Most of the fields were mined or, at least, had "Achtung! Minen!" signs in them. The hedgerows ran generally NNE-SSW and WNW-ESE, not north and south.

I found Sullivan. He asked me to locate our extreme left flank, find out the situation there, and report back. I worked my way to the left along the coastal road. In a couple of places I moved over to the south hedgerow and observed the firefight to my south. One thing stood out to me. Our rifles and machine guns put out a lot of smoke when firing. Try as I could, I was unable to detect any smoke or flash from the German rifles and machine guns engaging us. Bad show! The enemy could locate our weapons positions from the smoke they put out, but we were blind to theirs. I remember writing my father a week or so later about this smoke problem. He was the executive officer to the Chief of Army Ordnance.

I continued down the road to the east, looking for the end of our formation. In the process, I overran the flank a few yards and ran into a Ranger patrol just returning from the east. This patrol had made contact with and been joined by a 1st Division patrol (Stephen Ambrose insisted the patrol was from the 2nd Battalion, 116th Infantry and not the 1st Division), and both were returning to report to the 5th Rangers commander. They had a couple of paratroopers from the 101st Airborne Division with them, para-troopers who had been dropped into the water fifteen or twenty miles from their planned drop zone. After guiding the patrol back to Schneider and reporting to Sullivan, I dodged back and forth among the hedgerows until I got back to Headquarters Company.

ADVANCE ON VIERVILLE

C Company, 5th Rangers, had been providing support to E Company's attack with its 81-mm mortars. This fire was very effective, but because of its low volume, the Germans were able to reinforce their positions before E Company could advance.[62] The last envelopment was called back pending an attempt to get artillery support. The 58th Armored Field Artillery Battalion had gotten some guns ashore onto the beach, but the forward observer reported fire from the beach at this range was impossible because of the mask of the bluffs. About four hours had been consumed in these efforts. About 1400, Colonel Schneider gave up the attempt to move south from the highway and, instead, followed the coastal road into Vierville.

This route had been taken nearly four hours before by B Company, 5th Rangers and C Company, 116th Infantry. These units together with small groups of men from the rest of the 1st Battalion, 116th had passed through Vierville before noon and continued west on the coastal highway toward Pointe du Hoc. About five hundred yards beyond Vierville, they were stopped by well-prepared emplacements. During the next few hours, Company B, 5th Rangers and C Company, 116th Infantry worked together in efforts to outflank these positions. Here, they met the same kind of resistance and with much the same results as the remainder of the 5th Rangers had met in their attack south across the coastal road.

General Cota and the command group of the 116th

passed through Vierville about noon on its way to the pre-arranged CP location at the Chateau de Vaumicel. In Vierville, they found themselves in front of the rest of the friendly forces and uncomfortably isolated. This is when Lieutenant Gregory's platoon of B Company, 5th Rangers, came by on a flanking mission and was impressed as CP security. Small skirmishes took place near the CP all afternoon.

Around 1330, General Cota, leaving Colonel Canham in command, went down to the beach through the Vierville exit to see why no vehicles had come through. Naval gunfire had destroyed the German fortifications about 1300 and with that last resistance, the exit should have been open. Cota and his small group ran into nothing but scattered small-arms fire. They took five prisoners. All Cota found near the exit was the exhausted remnants of A Company, 116th Infantry and some tanks still trapped on the wrong side of the seawall. After prodding some engineers to start removing the obstacles in the Vierville draw, he walked the beach as far as the 1st Division sector beyond Les Moulins, a distance of over two thousand yards.[63]

About 1700, the main force of the 5th Ranger Battalion came up the road into Vierville.

At 1830, Lieutenant Colonel Metcalf, CO of the 1st Battalion, 116th reached the CP of the 116th Infantry. For the first time Colonel Canham, CO of the 116th Infantry, learned of the high casualties to the 1st Battalion, 116th, on Omaha Dog Green. He also learned his 2nd and 3rd Battalions were located near St. Laurent, some two and one half miles to the east.

The first vehicles, tanks of the 743rd Tank Battalion,

reached Vierville from the beach just before sunset, around 2200 hours.[64]

A COMPANY REACHES POINTE DU HOC

*A*t 1430 hours, Parker and his 1st Platoon left the chateau for the battalion assembly area, {63389092} southwest of Vierville. They used secondary roads, switching directions as the roads did in the hedgerows. The roads they followed ran between hedgerows that sometimes were head high, almost like a sunken road. On the way, they encountered a small enemy strongpoint, overwhelmed it, killing two and capturing twelve. They continued to the assembly area, found no one there, and concluded that the 5th Battalion had gone on toward Pointe du Hoc. Parker decided to follow.*

They ran into heavy opposition near Englesqueville. They could hear the Germans moving in the fields on both sides of them. When the enemy started to get behind them, they turned all their prisoners loose and retreated at the double time. Having gotten well beyond where the Germans were, they left the road and set off across country toward Pointe du Hoc.

About 2100 they contacted an inland group of the 2nd Rangers along the Vierville-Grandcamp Road. Word was sent to Rudder that Parker had arrived. Rudder's first question was where were the rest of 5th Rangers? Parker sent word back that the 5th Battalion must be close behind. His men were then incorporated into the 2nd Rangers defensive positions by placing them in small teams scattered throughout the area.

About midnight, the Germans, coming from the west and south, started conducting counterattacks, blowing whistles, using tracers. In several places and at several times, these attacks broke into the positions manned by the 2nd Rangers and Parker's men. The attacks continued until about 0200 the following morning. Then, without giving Parker any notice, the 2nd Rangers began to withdraw. When Parker finally learned of this, he and Lieutenant Zelepsky had the task of finding his scattered men in the dark in unknown terrain. Though difficult, they finally found them all and pulled back onto the Pointe itself.[65]

F COMPANY LANDS ON EASY GREEN

*T*he 1st Platoon, F Company, 5th Rangers, was landed in knee-deep water on Omaha Easy Green Beach at about 0900 hours on D-Day, by the same LCT that had plucked them from their sinking LCA. As the landing craft approached the beach, it was greeted by heavy artillery and machine gun fire. However, fire on the beach itself was light and no casualties were suffered in reaching the dune line. There, they found troops of other units packed in shoulder to shoulder. Patrols sent out to locate the remainder of the 5th Rangers were unable to gain contact. Moving west along the beach toward Vierville-sur-Mer, the platoon was hit by artillery fire taking eight casualties. After advancing six hundred yards to the west, the unit was engaged by a superior force and pinned down. When darkness fell the platoon stayed in its position.[66]

chapter 10

VIERVILLE-SUR-MER

1700

At 1700 the main Ranger force reached Vierville after severe fighting along the coastal road. Even though B Company, 5th Rangers, and C Company, 116th Infantry, had cleared this area a few hours before, the enemy had infiltrated back and reinforced his positions along the road, making progress slow and costly. After consolidating his positions and reorganizing his forces, Colonel Schneider developed plans for an attack to relieve the 2nd Rangers at Pointe du Hoc. However, Colonel Canham, now Schneider's immediate superior, called this attack off, realizing that to press this attack would deprive him of the very forces he needed to defend Vierville and the west flank of the beachhead. Canham then set up a defensive position around the west of Vierville. To the south he used the 1st Battalion, 116th Infantry, (minus) and elements of C Company, 21st Engineers. On the west he had the 5th Rangers, the Provisional Company of the 2nd Rangers, and C Company, 116th Infantry. After they arrived at 2200, the tanks of the 743rd were integrated into the defense positions around Vierville.

For me, the decision to move down the coastal road didn't change anything. We still had not set up a CP of any kind. Oh, sure, there was a command group with a couple of non-coms and radiomen always near Schneider, but not an operating CP. For Headquarters Company it was follow C Company. Stay in the ditches at the sides of the road. Stay down. Duck walk to keep that head low. Hit the dirt whenever the artillery began to come in. In training, I had always been a member of the command group, in on everything, knowing the situation, but not today. Today it was tag along, keep your men alive and ready for the moment we set up a CP.

Even when we reached Vierville, we still did not set up a real CP. Instead, my Headquarters unit was located in a farm enclave three or four hundred yards west of the crossroads and south of the road. {64229136} We were surrounded by farm buildings and walls—a more or less perfect place to defend. I organized it to provide as much defense in depth and all-round fire as possible. One of the line companies was to our front, while the command group was behind us in an adjacent farmyard. {64349130}

I sent three men down to the beach to see what they could scrounge. We needed food, water, and ammunition. The three came back with empty hands. "Captain, it was awful. For two thousand yards nothing but bodies lined up shoulder to shoulder. We just couldn't rob the dead."

Other patrols were sent out by the Ranger force. Graves was with Schneider and the command group. Someone called for seven volunteers for a patrol. Master Sergeant Harry J. Dunkle, the S-4 non-com, put his hand on Graves's shoulder and said, "We will go, won't we, Sergeant Graves," so Graves went as sort of a handcuff volunteer. Everyone in the patrol was a non-com, T-4 Victor H. Fast among them. The patrol found an old Frenchman with a cart and a horse. He agreed to go down to the beach with them. It was still daylight. The patrol made its way down the road to the beach. At the bottom of the road there was a large German pillbox that had been destroyed apparently by naval gunfire from the Texas. *All agreed, nothing else could have done so much damage.*

When they reached the beach Graves saw a terrible sight. There were dead men all over the beach. There were wounded men in terrible trouble. Some had lain there all day and were out of their heads. There were men crying, men moaning, and there were men screaming. The wounded and the dead were everywhere, just a great mass of them. The few medics on hand were scurrying around trying to get to as many of the wounded as they could. To Graves, it was the most terrible thing he would see all through the war. It looked as close to how he imagined Hell.

The patrol managed to get the ammunition they came for by prowling around the dead and dying. They concentrated on 60-mm mortar and machine gun ammunition. They got their load onto the cart and made their way back to the headquarters area. The companies came in later to resupply themselves.[67]

With the cliffs at Pointe et Raz de la Percée and the beach exits firmly under control, the remaining twenty-eight men of C Company, 2nd Rangers (Force B) had finally been relieved by other troops. They set out through the Vierville draw where they met a 5th Ranger patrol, probably the one from Headquarters that had returned to the beach in a search for ammunition. C Company was then led back to the 5th Rangers, arriving about 2200 hours. Apparently, Lieutenant Salomon's platoon did not follow the rest of C Company, but spent the night in the fields near the stone house.

After organizing our defense and setting the guard, I ordered the men to dig in. At that moment I found out my worst mistake of the day. I had lost my entrenching tool or left it aboard the *Prince Baudouin*! A couple of the men volunteered to dig a hole for me, but no, I would have none of that. I tried to use my helmet as a scraper, but centuries of hard-packed clay yielded only a scratch or two. Night was falling and I had to do something. I spied a haystack, more or less in the middle of the courtyard where I had been trying to dig. A haystack would be warm, the weather seemed like it was nearly freezing. The hay would slow down fragments a little bit, so why not. I scooped away a little hay, lay down and covered up. It was warm.

Now, I am a city boy, so I didn't know the difference between a haystack and a manure pile. I learned that difference in seconds. Every kind of biting insect in France was at me, hundreds of them. I came out like a shot. In between their laughter, some of the men came up with insect powder, which I liberally applied under my clothing. The powder slowed down the biting, but didn't do a thing for the itching.

I saw a flickering light in a building a short way down a path. I went there more to enforce blackout than for any other reason. I found Lieutenant Van Riper sitting on a wooden bench watching an old French woman making a fire from faggots. The tiny fire was the source of the light. After covering the window and notifying the first sergeant where we were, Van Riper and I spent most of the night in that hut, trying as best we could to keep warm by the tiny fire without interfering too much with the woman's cooking.

Thus ended my "longest day." Ignominiously.

chapter 11

ST. PIERRE DU MONT

———

The story I wanted to tell stops at the end of June 6, 1944. However, I would feel remiss if I didn't at least tell you how we relieved the 2nd Rangers at Pointe du Hoc. After all, that was our mission.

During the night of D-Day, Colonel Canham, Lieutenant Colonel Metcalf, Canham's 1st Battalion commander, and Lieutenant Colonel Schneider met with their staffs in the farmhouse where the 5th Rangers headquarters was located. During those meetings, we made plans for the relief of the Ranger Force A at Pointe du Hoc, the capture of Grandcamp-les-Bains, and the push towards Isigny.

The plan that developed was for the 116th Infantry, supported by tanks of the 743rd Tank Battalion and with the 5th Rangers attached, to drive straight for the Pointe. The Provisional Company of the 2nd Rangers supported by eight tanks of the 743rd Tank Battalion would serve as the advance party. Elements of the 1st Battalion, 116th Infantry would come next. This was primarily C Company under Captain Bert Hawks, followed by the 5th Ranger Battalion as the Advance Guard Support Force. The main body would consist of the 2nd and 3rd Battalions of the 116th Infantry, the remaining tanks of the 743rd Tank Battalion, and attached divisional units.

For me, D+1 started with a bang. The Germans attacked from the south with what seemed to be company strength. With the help of the 743rd Tank Battalion, parked on the main east-west road through Vierville, we held them off. The tanks had come up during the night, and this fight was the first I had seen of them. The tank crews were all buttoned up and could not see or hear that we were being attacked. I jumped up on the hull of one and banged my rifle butt on the turret until a tanker opened his hatch. I then pointed out the attack and suggested he take the enemy under fire. He and the other tanks did just this, but only fired their caliber .50 machine guns, apparently believing that the targets were too undefined for them to use their 75-mm cannons.

Right after the German attack fizzled out, Sullivan had me take a patrol out to our north and east. I had Minor Dean, the battalion sergeant major, and Harry Dunkle, the battalion supply sergeant, and Corporal Jack Sharp with me. We crossed the road going north from the headquarters into the same field that Lee Brown, my company clerk, had tripped a "bouncing Betty" anti-personnel mine the night before. The anti-personnel mine was a dud. Needless to say, we were very, very careful. We turned right (east) and cleared the area south of the main road of any enemy stragglers.

We uprooted two and had little firefight with them before they broke and ran across the road to the beach and disappeared behind the military crest. We pursued them past the crest but could only find a feldgrau field hat at their last position. From the sound of tank engines revving, it was clear that the relief force headed for Pointe du Hoc was getting under way. Dean and Dunkle felt their place was with that force and left Sharp and me while they rejoined the battalion. I, on the other hand, had been told to locate the 29th Division CP and report our situation and plans there.

Sharp and I went back to the road running to the beach, turned right (north), and headed toward the shore. It wasn't long before we ran into an MP traffic control point. The MPs directed us to the Divisional CP, where we were finally admitted after some severe questioning. General Charles Gerhardt had apparently heard we were there and came out and invited us into his operations center. The CP was located in an old quarry near to the forts guarding the Vierville exit. It was pretty rudimentary. After I reported the situation and plans of the 5th Rangers to him, General Gerhardt asked if there was anything he could do to help the Rangers. I told him we needed ammunition badly, and he gave me permission to take any abandoned vehicle that we, Corporal Jack Sharp and I, could find. We picked up an abandoned jeep, still waterproofed, belonging to the S-4 of the 58th Armored Field Artillery Battalion. With the help of a sergeant from the Engineer Beach Brigade, we got it running. General Gerhardt's aide-de-camp, who had accompanied us to the beach, had some 29-ers at the temporary ASP (Ammunition Supply Point) on the beach fill it up with mortar and machine gun ammunition. Sharp and I took off up the road to Vierville.

While I was on that patrol, plans for the D+1 operations were changed and this is how that happened.

The march of Colonel Canham's force, with Schneider's Ranger force attached, started between 0730 and 0800 as planned. The movement of the column was well under way with the advance party and most of the support party clearing the IP (Initial Point). Suddenly, the enemy counter-attacked from the south. Lieutenant Colonel Schneider had a hurried decision to make. His C and D Companies and the Provisional Company of 2nd Rangers were already on the road to Pointe du Hoc. He could have recalled them but decided the relief force needed their presence to be an effective force when they got to Pointe du Hoc. He did hold back B and E Companies and the platoons from A and F Companies of the 5th Rangers to protect Vierville.[68]

Schneider's actions left a small battalion-sized force on the road to relieve Pointe du Hoc. This force was under Lieutenant Colonel Metcalf and Major Sullivan. It included the Provisional Company of the 2nd Rangers with Rangers from A Company leading, eight tanks of the 743rd, C Company, and other remnants of the 1st Battalion, 116th and C and D Companies of the 5th Rangers.

The town of Vierville appeared empty as Sharp and I came up from the beach and drove through. The column under Metcalf and Sullivan had long since moved out toward Pointe du Hoc. Mostly on faith, Sharp and I sped down the road after them. It was a very sporty trip. Four different times, snipers fired on us, one bullet hitting my helmet and spinning it into my lap. At one point, the machine gun and sniper fire were so intense that Sharp and I got out of the jeep, slipped under it, and pushed it over our heads past a break in the hedgerows while bullets smacked into the ammunition boxes above us. We had serious doubts when we reached a fork in the road {63149226} just north of Gruchy. Which way, right or left? I dismounted and examined the battle debris and found lots of tank tracks, cartridge brass, machine gun links, and fiber containers down each road of the fork. Luckily I chose the right, and we soon caught up with the column, distributed the ammunition, and gave some of the walking wounded a much-needed ride. When it ran out of gas, probably from a bullet-induced leak, we abandoned the jeep in a farmyard north of the road {60469296}, just short of St. Pierre du Mont {59869286}.

THE RELIEF FORCE ADVANCES[69]

*T*he advance of Metcalf and Sullivan's relief column toward the Pointe was, under the circumstances, very rapid. The Provisional Company of the 2nd Rangers took the point, with its scouts well out in front, followed by a double file moving on each side of the highway. The tanks were positioned just behind these 2nd Rangers but moved freely up and down the column as needed. Many enemy positions were encountered along the line of march, but none were strong enough to slow down the point, let alone stop it. If fired on, the point simply pushed ahead and the tanks would come up and suppress the enemy position as they apparently had done at Gruchy. German prisoners *later commented on the "crazy march" past their prepared positions just off the road.*

By 1100 hours, the main body of the relief force had reached St. Pierre du Mont and stopped fifteen hundred plus yards from Pointe du Hoc. The advance party had run into difficulties.

Two hundred yards beyond St. Pierre du Mont, the road curves to the south near the chateau {59049297}. A large crater with marked minefields right and left, blocked the tanks. Taylor, in "War Department Notes," wrote that naval gunfire had made the crater. When I came on it later, I thought it was a prepared demolition flanked by minefields to make it more effective as a road block. The crater stretched across the entire road and was perhaps fifteen feet deep.

Captain Arnold, B Company, 2nd Rangers and in command of the Provisional Company, sent back for the tank dozer. The rest of his company stopped. However, the point kept going forward until it was fired on by a machine gun near Au Guay {58759300} and could see Germans in the village houses. The scouts found themselves pinned down. Sergeant White took six men forward to rescue them. White was able to knock out the machine gun, and all of the advance party were able to move into Au Guay, about two hundred yards from the exit road leading from the Pointe du Hoc.

By now, a tank had gotten past the roadblock, and it came up in time to help with the machine gun. Other tanks now arrived, but enemy mortars and artillery began hitting all along the road east of Au Guay to the chateau some five hundred yards away. The entire area north of the road was wooded, and this interdictory fire, probably high angle, 10- or 15-cm howitzers, yielded many tree bursts. The artillery fire stopped the tanks and forced them to withdraw. The advance party at Au Guay was ordered to withdraw since their tank support was gone. As they moved back toward the crater, they met C Company of the 116th going forward. The Provisional Company withdrew to just west of the crater, west of St. Pierre du Mont.

C Company of the 116th, supported by D Company, 5th Rangers and tanks of the 743rd now pushed forward. The tanks reached Au Guay, but the following infantry was caught by a very heavy artillery concentration short of Au Guay and, after suffering heavy losses, withdrew in disorder to the crater.

The tanks continued down Route D 514, the coastal highway, and pushed on past the exit road of Pointe du Hoc, past its infantry, and got to a point {vic 58159313} where the lead tank crew could see Grandcamp. Then, with no infantry to support them, the tanks withdrew back through Au Guay and into St. Pierre du Mont.

Having abandoned the jeep in Le Fevre. I continued forward on foot and found Sullivan at the edge of the village. I learned that Schneider and most of the 5th Rangers were still back in Vierville and that he, Sullivan, and the other Rangers with him were part of Metcalf's relief force. Everyone in the St. Pierre force from Metcalf down was concerned about rumors that the German counterattack earlier in the morning had been successful and Metcalf's force was now cut off. My getting through partially reassured them that the beachhead had held. Even so, Sullivan asked me to check out the tanks to see if they were ready for a defensive battle for St. Pierre du Mont. I found several tanks in good defensive positions throughout the hamlet and then got to the crater. I was told that the marked minefields to the right and left of the crater had blocked the tanks and that none had crossed the crater. This was wrong information and at least four tanks had gotten past the crater and were even then involved in actions around Au Guay.

While I was standing there on the south edge of the crater, heavy artillery began falling perhaps two hundred yards in front of me. A few moments later infantry began streaming by me through the crater.

I saw that many were Rangers and used many a choice word at them. I was standing on the rim of the crater and the troops were down in the crater, well below me. The fire that had driven them back didn't fall in the vicinity of the crater, so there was nothing to drive them now but panic. By the time those in front had crossed the crater and come up on the other (east) side, the panic had subsided and they became more orderly. Initially, the men were from the 2nd Rangers, but now as Rangers continued by, I began to recognize some of them and was able to call them by name. These were now 5th Rangers. When they heard their names intermingled with my curses, the men hauled up, particularly when I yelled at the officers. I grabbed the company officers, got them to reorganize and get back across the crater into battle positions.

Sullivan appeared and called me over. The 29th Division

people were worried about the security of the beachhead and wanted all the tanks they could find to help with its defense. As a result, he, Metcalf, and the tanks were pulling back to Vierville for the night, leaving C and D of the 5th Rangers, the Provisional Company of the 2nd Rangers, and our inevitable comrades, C Company of the 116th to hold the positions around St. Pierre and that I was in command. "The monkey's on your back," was the way Sully put it.

As the tanks left us, I called off the third attack toward Au Guay and pulled the infantry back into positions surrounding St. Pierre du Mont.

THE DEFENSE OF
ST. PIERRE DU MONT[70]

Communications had been poor on D-Day and were still unsatisfactory. Metcalf's force had no definite word as to the state of affairs at the Pointe and watched anxiously in that direction all afternoon. What they observed led them to conclude that the Ranger force at the Pointe had left or had been wiped out, and a naval gunfire mission on the Pointe was requested. The navy's answer that they were firing missions in support of the Rangers at the Pointe gave the relieving force their first definite assurance that the Pointe was still held. Although the Rangers on the Pointe were engaged all day in intermittent firefights, involving much mortar and naval gunfire support, the relieving force could not identify that action in the general noise and confusion of battle.

Lack of communications gives a good background for the development of battlefield rumors. After the failure of the afternoon attacks, the enemy showed signs of activity that might indicate a counterattack on St. Pierre du Mont. The Rangers had been briefed before the invasion to expect enemy counterattacks in battalion strength on D+1, and even for use of enemy armor.

With the tanks gone, Wise and I met and were discussing the situation. As we did, a 29th Division field grade officer came up to us with the news that Vierville had been retaken by the Germans, the beachhead liquidated on that flank, and enemy armor was coming up the road from Vierville on our rear!

This information spread like wildfire among the men, and their morale dropped off for the first time. Wise and I found the information perfectly credible and discussed our options, the first of which was a last-ditch stand at St. Pierre du Mont. We even admitted to each other that it was doubtful that we could hold the village if it were true that we were about to be attacked by armor.

We flipped a coin to see which one of us would go back to Vierville and reconnoiter the enemy strength and movements. I won the toss and set out on this "desperate" mission. I hadn't gotten five hundred yards when I met a 29th Division officer riding a bicycle west and looking very calm about the situation. I stopped him and asked for information on the situation at the beachhead. He told me that the 175th Infantry Regiment had just landed and was moving through Vierville towards us. The 175th was headed by armor and *would reach us in a few hours. That story, although wrong about the route of the 175th, restored our confidence, and for the first time we learned that "the invasion was a success."*

About 1710, the 2nd Rangers on the Pointe contacted Schneider's headquarters in Vierville using an SCR 300 radio. The 2nd Rangers asked if contact could be made that night between the Pointe forces and the relief force. The reply was given that this would not be possible.

That night, sitting on a dirt floor in a barn, sketching the terrain and positions in the dirt with twigs, Wise, Hawks, and I reworked our defensive positions. At one point, we all burst out laughing at the ridiculousness of our situation. After dark we sent Sergeant Moody and Corporal McKissick, both from my old platoon in C Company, on a two-man patrol to locate the main force of the 2nd Rangers. I knew both these Rangers well and was delighted when Wise chose them for this difficult and dangerous mission. And so I want to sleep, this time in a hayloft, deep in enemy territory, surrounded by the enemy.

F COMPANY'S PLATOON REACHES POINTE DU HOC[71]

"The 1st Platoon of F Company, 5th Rangers, that had been landed near St. Laurent-sur-Mer attacked inland from their overnight positions on the beach at 0800 hours on D+1. By 1400 hours they had secured their objective, capturing three pill boxes and several weapons emplacements."*

Throughout the day Lieutenant Reville had inquired of the Beach Brigade's Engineers for any news of the 5th Rangers. That paid off, for the platoon was contacted by Major Jack B. Street of Admiral Hall's staff, loaded into an LCVP [Lieutenant Ike Eikner, the communications officer of the 2nd Rangers reports that this was an LCT that earlier had stopped at Pointe du Hoc to deliver supplies], and transported to a destroyer lying off shore, probably the USS Harding. [Another version of the story says the ship was the USS Texas, not the Harding.] Street had been a company commander in the 1st Rangers in Sicily, and learning of the dire straits of the 2nd Battalion at Pointe du Hoc did everything he could to get them help. Aboard the Harding, the platoon was fed, wounds tended to, and loaded into another LCVP with food, water, and ammunition and then transported to Pointe du Hoc. The Rangers griped the whole way up the cliffs, not about going back into combat, but having to carry the water and other supplies up the cliffs like a bunch of stevedores.

Runge and Reville reported to Colonel Rudder at about 1700 hours on D+1. The platoon was integrated into the defensive positions of the 2nd Rangers. About 2300 hours, an eight-man patrol, headed by Lieutenant Reville, Company F, 5th Rangers, was sent out to locate the 5th Rangers. As it proceeded down a path between hedgerows in almost total darkness, Reville was brought sharply to his senses when he heard, "HALT! THROAT!" For a moment he could not recall the countersign for "Throat." Then it came, "WEAPON!" The patrol had stumbled upon a machine gun outpost *of C Company of the 116th. They were led back to my CP at St. Pierre du Mont, arriving at 0800 hours of D+2.*[72]

5TH RANGER ACTIONS IN AND AROUND VIERVILLE[73]

"*Elements of four companies of the 5th Ranger Battalion (Companies B, E, and 1st Platoon of each A and F) had been stopped when the column moved off in the morning toward Pointe du Hoc. These units were given the mission of protecting Vierville against reported German counterattacks, so that movement from D-1 (the Vierville exit) through the village could continue.*

"*Captain Luther was ordered to take his E Company through the village to clean out snipers, who had reappeared. Luther organized his company into six groups, each responsible for a part of the village, they searched every house, but found only a couple of snipers. Two houses from which firing had come were burned. This job took all morning. In the afternoon, Luther was ordered by Colonel Schneider to set up a defense along the southeast side of the town, toward the Chateau area.*

"*The platoons of A and F operated together (about sixty-five men). They went into Vierville with elements of two tank companies of the 743rd Tank Battalion and turned south at the main crossroads toward the chateau. The enemy counterattack which had caused the alarm had come from this direction, but had long since receded. The Rangers found only wounded engineers at the chateau, met no en-*

emy fire, and returned to Vierville. They made three more trips between Vierville and the chateau during the day, never encountering enemy in any strength.

"During the afternoon, A Company occupied the chateau for several hours and searched it thoroughly. Ten German stragglers were picked up in the brush nearby. Machine gun fire from German weapons to the south caused a moment's flurry, and a few mortar shells were fired before it was found that friendly troops to the south were using German weapons. Late in the day units of the 175th Infantry passed the chateau on their way south. Between 1800 and 1900 hours, enemy artillery from the south began to shell the vicinity: figuring the chateau was probably zeroed in, the two platoons drew back into Vierville and then took positions west of the village for the night.

"This withdrawal from the chateau had a repercussion on E Company. Captain Luther stayed on the southern edge of Vierville during the afternoon, protecting an ammunition dump and awaiting relief by the 3rd Battalion, 116th Infantry. When the enemy artillery opened, his company had five casualties. Shortly before 1900 hours, an enemy attack in company strength came straight toward Vierville from the chateau grounds, recently left unoccupied. Luther's men had scattered during the artillery fire, and he was able to get about twenty-five together along an east-west hedgerow. He had no contacts on either flank. Forty rounds of [60-mm] mortar fire were placed on the wall at the north edge of the chateau grounds, where the enemy was assembling. This seemed to check the attack; Luther saw some enemy close

up on his right flank, but they disappeared to the west. In half an hour, the firing died down, the 3rd Battalion, 116th came up, and the last threat to Vierville was over.

"Luther moved E Company west through the village. On the way, naval fire began to hit just beyond his route, and he had to put up yellow smoke to check it. His company was put into an outpost position four hundred yards west of Vierville and had a skirmish with enemy patrols evidently trying to out post the same fields.

"B Company had spent the day in this area, with only occasional patrol activities against snipers. The four companies (A, B, E, and F) of the 5th Rangers held positions during the night to protect Vierville from the west and southwest."

Shortly after 0600 on D+2, the portion of the 5th Ranger Battalion left in Vierville assembled and moved out in the lead of the 116th Infantry's advance on St. Pierre du Mont.

chapter 12

RELIEF OF POINTE DU HOC AND THE REST

D+2 dawned and I learned Moody and McKissick were back. Hurrying to get their report I asked, "Did you contact the 2nd?" "Yes sir." "Any messages? What shape are they in?" At that Sergeant Moody handed me a field phone and said, "You can ask them, Captain." Those two Rangers had worked their way through 1000 yards of enemy territory, gone too far west and overshot Colonel Rudder's CP. They then doubled back, found the CP and gave Colonel Rudder the location of all Ranger units and the story of the 175th coming up from Vierville. Colonel Rudder told them he could hold out till the next day. The patrol returned to St. Pierre du Mont early next morning, laying field wire from La Pointe du Hoc back to our positions in St. Pierre du Mont. All this during several German counter-attacks on the Rangers holding the Pointe. At this time, there were nearly one hundred

2nd Rangers and fifty 5th Rangers holding the Pointe.

I had planned to attack from our positions at St. Pierre du Mont towards the Pointe as soon as we could disengage from our night defensive positions, but as we did so, about 0800, the main force from Vierville arrived in St. Pierre telling us of the six battalion attack on Pointe du Hoc that had been planned during the night. In this effort, Moody and McKissick's phone proved invaluable as Colonel Schneider was able to tell Colonel Rudder of the plans for the attack. These plans were that the Ranger force (less Companies B and E, 5th Rangers) and the 1st Battalion, 116th would attack directly at Pointe du Hoc. Companies B and E, 5th Rangers would seize and hold the high ground west of the sluice gate at Grandcamp-les-Bains. {56329326} The 2nd and 3rd Battalions, 116th and the 743rd would continue their attack down the road toward Grandcamp-les-Bains. When they were west of Pointe du Hoc, the 3rd Battalion, 116th Infantry and its attached tanks would wheel to the right enveloping any Germans still in the area. The 2nd Battalion and its tanks would follow Companies B and E of the 5th Rangers to attack Grandcamp.

Everything went according to plan. Our force jumped off at 0900 and advanced on Pointe du Hoc with only light resistance. By mid morning, 1000 hours, we had relieved the 2nd Rangers and were sitting around on the rubble of the gun positions swapping our stories, when all of a sudden tanks of the 743rd Tank Battalion burst out of a patch of woods directly to our south and attacked us with machine gun and cannon fire. A lieutenant of the 2nd Rangers, ran out of cover, jumped up on one of the tanks, beat on the turret until he got the attention of the crew, put his pistol to the head of the tank commander who had opened the turret and with that the attack by the 743rd stopped in its tracks. 5th Ranger casualties from that attack were two killed and four wounded.

By noon, it was all over with Pointe du Hoc essentially cleared of enemy.

THE CAPTURE OF GRANDCAMP-LES-BAINS

Meanwhile Companies B and E of the 5th Rangers led the 116th Infantry Regiment and the tanks of the 743rd down the coastal road towards Grandcamp-les-Bains. The mission of B and E Companies of the 5th Rangers was to take and hold the high ground west of the sluice gate at Grandcamp-les-Bains. At 1000 with B Company leading, the force moved into the outskirts of Grandcamp-les-Bains. Initially, no fire was received, but as they reached the bridge leading into the town, they received heavy mortar and machine gun fire. The two companies withdrew to the high ground east of town where they were reinforced by D Company, 5th Rangers, which had just come from Pointe du Hoc.

A battalion of the 116th Infantry passed through the three Ranger Companies and with the support of the tanks, artillery and naval gunfire captured the town.

After some minor reorganizing, Companies A, C, F, and a Headquarters element, 5th Ranger Infantry Battalion, proceeded from Pointe du Hoc to the southern edges of Grandcamp-les-Bains under Major Sullivan. A Company provided left flank protection to Sullivan's force and the 116th Infantry. In A Company's area of advance, the Germans had flooded the low-lying land so that A Company often had to move in knee deep water in the area south of the coastal highway.

My headquarters element was attached to Major Sullivan and given the mission of clearing the houses along the road through Grandcamp of any German stragglers. I set up four teams of four Rangers, two for the right side of the road and two for the left. I led the first team on the left. The team would enter a house. If the door were locked, a Ranger would stick a bayonet in the keyhole, fire one round, and kick open the door. It works! First man up the stairs was to clean out the second floor. The second man was his backup. Third and fourth man were to handle the main floor. Then the first two finished would take the basement. The teams would leapfrog from house to house. Luckily, we found nothing but frightened French families for our efforts.

Back in the countryside, we worked our way toward Maisy, halting about a half mile to the northeast {vic 54009280} where we snuggled into the ditches and hedgerows for the night. By now, the battalion trains had at least given us K Rations to eat rather than chocolate D Ration bars.

Meanwhile, D Company had been detached from its mission at the sluice gate and ordered to make a sweep through Criqueville and back into Maisy. It "bivouacked" overnight near the chateau at {54339190}.

THE ATTACK ON THE MAISY BATTERY

On the morning of D+3, June 9, 1944, A, C, and F of the 5th Rangers, still under Major Sullivan, were detached from the 1st Battalion, 116th Infantry. In its advance to the south, the 116th Infantry had encountered severe resistance from German strongpoints at La Martiniere {vic 52759150}, three quarters of a mile, SSW of Maisy and Les Perruques, {vic 53309180}, a half mile SW of Maisy. On Map 4, La Martiniere is located at artillery registration point "16" and Les Perruques at point "5." Together these two connected positions were called the "Maisy Battery." The 116th's objective was Isigny not Maisy, so the 116th Infantry, after calling in some naval gunfire, bypassed these German positions leaving them for the 5th Rangers.

There was a third artillery battery position in the Maisy complex. Its target number was "108." It was located {vic 53059130}. It contained 150-mm field howitzers. It was destroyed on June 8th by the USS Shoebrick, a destroyer, using an artillery spotter airplane and high register fire. It seems the position was located on the back slope of a hill, and low register fire fell short into the hill or long into the valley beyond. This battery played no part in the battle on June 9, 1944.

The Maisy Battery was quite a formidable fortification. It consisted of three batteries of artillery, extensive minefields, a major communications center and a large medical complex. There were, of course, many troops to defend such an important fortification. These troops included Army, SS and Luftwaffe personnel.

Les Perruques was home to a six-gun battery of 155-mm Schneider World War I howitzers. Four of these guns were in concrete pits and two were in open emplacements. Since the howitzers had a maximum range of a little more than 12,000 yards, about a half mile short of the D-1 or Vierville Exit, the howitzers appear to have had a principal mission of defending the Vire River valley from amphibious assault rather than defending

Omaha Beach. In all probability, the Maisy 155-mm howitzers were used against the 2nd Rangers on Pointe du Hoc and against Metcalf's force as it attacked through St. Pierre du Mont on D+ 1. La Martiniere had four 10-cm Skoda Field Howitzers. These guns had been rebored to the standard German 105-mm caliber to simplify ammunition resupply logistics. Three of these guns were in casemates and one was in an open pit with the casemate for it under construction.

The German defensive positions from Grandcamp and Maisy westward along the coast and up the Vire valley showed a genuine fear that an Allied attack could come up the valley. From Grandcamp westward there were 20 Widerstandsnester (WNs), small resistance points manned by a squad or more, plus supporting minefields. On the western side of the valley in the VIIth Corps area there were an additional six such WNs, blocking entry into the Vire valley from the Utah Beach side as well.

Les Perruques, WN 83, and La Martiniere, WN 84, were two of the more gigantic resistance points, gigantic because they contained artillery and other facilities as well as infantry fighting positions.

The minefields protecting the Maisy complex were extensive. While the position itself was about 1,200 yards by 500 yards, the minefields were 1,200 yards by more than 1,000 yards. The mines outside the perimeter of the Maisy fortifications were oriented to block an attack from the northwest, the Vire River Valley. Most of the mines were anti-personnel, "Bouncing Bettys." In the area where A Company would attack, these mines were connected together in sets of about ten. When one mine of the ten was tripped the others would also detonate. And that happened when Pfc. John Bellows of A Company tripped a mine. Zipkac, Siemens, Baptist, and Burke were wounded. The anti-tank mines needed to complete the minefield had not been delivered by D-Day. The entire area was clearly marked with "Achtung Minen" signs.

As he approached the fortifications, Major Sullivan had three 5th Ranger Companies, two 75-mm gun half-tracks from the 2nd Ranger Battalion, and Company B, 81st Chemical Weapons Battalion. The latter unit was armed with 4.2-inch mortars. Major Sullivan made his headquarters in one of the 2nd Rangers' half-tracks. During the approach march, I was at the rear end of the column with a small headquarters element. Crossing through the hedgerows and fields we were taken under long range machine gun fire several times. However, we were beyond tracer burn-out and the Germans were never able to adjust their fire well enough to even bother us.

Because of the extensive minefields protecting the German strongpoint at Maisy, Major Sullivan decided to approach the position from the north, initially in a column of companies with F Company in the lead followed by A Company with C Company in reserve. Sullivan had an artillery liaison officer with him and put him to good work. The 58th Armored Field Artillery Battalion bombarded the Maisy positions in preparation for the 5th Ranger advance. Support during the attack was provided by the 81st's 4.2-inch mortars, the two 75-mm cannon mounted on the 2nd Ranger's half-tracks and the four 81-mm mortars of C Company of the 5th Rangers.

Company F moved down the dirt road just north of gridline 92. About 0800 hours, the company came under fire near {53459220}. The company deployed, took cover and began an exchange of scattered rifle with neither side accomplishing much.

A Company then passed by F Company on its right in its approach to its line of departure for an attack on La Martiniere. The area to the left of the path A Company was on was quite swampy, and in this swampy area, A Company came across some dead American paratroopers who had been dropped miles from their intended drop zones. Some had drowned, some had been caught up in the trees and gunned down by the German defenders.[74]

Four paratroopers were captured by the Germans early on D-Day between Maisy and Gefosse-Fontenay and brought to Maisy. Seventeen additional paratroopers were captured in later on D-Day actions. These paratroopers were placed in one of the hospitals on the south edge of La Martiniere and later evacuated before the A Co attack.[75]

When opposite La Martiniere, vic {52769200}, A Company heard heavy firing as F Company began its attack on Les Perruques. A Company took this as a signal to begin its attack on La Martiniere. Captain Parker ordered the company to "Fix Bayonets!" The company immediately wheeled to the left and attacked across a boggy area. The company's initial attack bogged down in deep water as it crossed a flooded area, and it was necessary to pull back and move more to the west for a second attack. A Company's second attack progressed well, penetrating the La Martiniere *position vic {52579170}, with some of the German defenders laying down their weapons and surrendering. However, some German officers, possibly SS, began shouting threats and shooting some of their own men in the back. From there on, the defense stiffened and nobody dared surrender.[76]*

Abandoning the firefight it had been in, F Company crossed the boggy area and began an attack straight down the line of outer fortifications, starting near {53359200}. This outer line consisted of seven tobruks, or machine gun pill boxes, spaced about two hundred yards apart. Lieutenant Reville's platoon smothered the tobruk with small-arms fire, then moved several Rangers close enough to throw grenades and satchel charges into the open top pit. Three tobruks were reduced this way, by which time the enemy was demoralized.[77]

Meanwhile, other things were happening. A Company began to receive heavy rifle and machine gun fire from the south and east. Its attack was slowed as it had to slug its way through the position. Dr. Petrick was with A Company during this attack. He set up an aid station right behind the attacking forces and impressed a German doctor he found to assist in treating the wounded of both sides.[78]

C Company and the cannon platoon had turned down the southerly road from Maisy to about {53509170} and swung to the right, attacking Les Perruques from the southeast. C Company's mortars were left in position near that same point. The self-propelled guns supported C Company's attack with direct fire on targets of opportunity. During this attack, Major Sullivan dismounted from his half-track to get closer control over the battle. He was wounded, though not seriously, when a nearby mine detonated. As C Company and F Company began to interfere with each other's operations, Sullivan ordered C Company to withdraw. After moving southwest, C Company turned to its right and attacked northeast from vic {53049132} through the forces giving A Company so much resistance.[79]

Despite the minefields and stubborn resistance, the Maisy Battery was successfully captured in a five-hour battle by this three-pronged attack. A fairly large number of D Company Rangers were involved in the attack on Maisy, scattered throughout the attacking companies. As the medic attached to A Company, Jack Burke, put it, "At that point we were often bundled together by anyone—especially if people couldn't find the exact location of their own unit."

After the battle, the two half-tracks were sent back to the beach with the wounded. La Martiniere contained three 10-cm howitzers (the fourth gun was probably destroyed by bombing and/or naval gunfire), large stocks of ammunition and other supplies, and about ninety defenders who became POWs.

The Les Perruques position contained six 155-mm Schneider howitzers, a major headquarters complex, and about fifty prisoners. The guns in the "108" position had been destroyed by naval gunfire before the attack began. The POWs and the position were turned over to elements of the 29th division. One hundred and eighty tons of ammunition were removed from the Maisy site by Engineer units after the battle.

Ace Parker, the A Company commander, later told me that, as

far as he was concerned, the fight at Maisy was far worse than the Omaha Beach landings of four days earlier.

After capturing the Maisy Battery, Sullivan's Ranger force marched to a bivouac area west of Osmanville where, at 2000 hours, it rejoined the rest of the battalion and turned in for the night in the hedgerows and ditches. The 2nd Ranger Battalion also spent the night in the same area. All was not quiet in Osmanville. "Bed Check Charley," a German airplane pilot, paid nightly visits to the battalion area, occasionally dropping bombs. Several men were wounded in these nightly forays.

The next two days were spent in mopping up the coastal fortifications from Grandcamp-les-Bains to Isigny and in patrols in the vicinity of the battalion bivouac area.

In five days of fighting, the battalion suffered twenty-three men killed, eighty-nine wounded, and two missing. Approximately 850 prisoners were taken and some 350 Germans killed.[80]

ENDNOTES

1. After Action Report, HMS *Baudouin.*
2. Graves, James W., Jr., conversations and correspondence.
3. Parker, Charles H. *Reflections of Courage on D-Day:As told to Marcia Moen and Margo Heinen,* DeForest Press, Elk River, MN. Also, Conversations and Correspondence and an audio tape.
4. Black, Robert W. *Rangers in World War II* (New York: Ballantine, 1992).
5. After Action Report, 507th Flotilla.
6. After Action Report, USS *Texas.*
7. After Action Report, 504th Flotilla.
8. After Action Report, HMS *Prince Baudouin.*
9. Hastings, Max, Sir. *Overlord: D-Day and the Battle for Normandy* (New York: Simon and Schuster, 1984).
10. Graves.
11. War Department Notes.
12. After Action Report, HMS *Prince Baudouin.*
13. *Ibid.*
14. Balkoski, Joseph. *Omaha Beach, D-Day, June 6, 1944* (Mechanicsburg, Penn.: Stackpole Books, 2004), 50.

15. Herring, Thomas F., conversations and correspondence.

16. *Small Unit Actions*. War Department, Historical Division, Washington, D.C.

17. *Omaha Beachhead*. War Department, Historical Division, Washington, D.C.

18. Ibid.

19. Ibid., 11, 20, 35, 37.

20. After Action Report, ML 304. See copy in Appendix H.

21. *Omaha Beachhead*, 88.

22. *Small Unit Actions*, 7.

23. Wells, Theodore H., conversations and correspondence.

24. *Omaha Beachhead*, 30, 42.

25. Hicken, Victor, Ensign, U.S. Navy, OIC, LCT(A)-2227, correspondence.

26. Merrill, Richard P., conversations and correspondence.

27. Graves.

28. *War Department Notes*.

29. Ibid.

30. *Omaha Beachhead*, 31, 53ff.

31. Graves.

32. *War Department Notes*, and Parker.

33. Parker.

34. *War Department Notes*.

35. Ibid.

36. Bliven, Bruce, Jr. *The Story of D-Day, June 6, 1944* (New York: Random House, 1994), chapter 13.

37. *Omaha Beachhead*, 13.

38. *Omaha Beachhead*, 58.

39. *War Department Notes*.

40. *War Department Notes*, and Parker.

41. Herring.

42. Graves.

43. Ibid.

44. Shae, Jack, 1st Lieut., Inf., aide-de-camp to Brigadier General Norman Cota. *Extracts from Combat Narrative, Second Draft*.

45. Reed, Elias E., Jr., conversations.

46. *War Department Notes* and conversations with Captain Edward S. Luther.

47. *War Department Notes*.

48. Reed.

49. McIlwain, Walter N., correspondence and conversations.

50. Ibid.

51. *War Department Notes.*

52. Ibid.

53. Ibid.

54. Ibid.

55. Parker.

56. Bryant, Stewart, correspondence and conversations.

57. Parker.

58. Marshall, S. L. A., "First Wave at Omaha Beach," *The Atlantic Online*, November 1960.

59. *War Department Notes.*

60. Glassman, Henry S. *Lead the Way, Rangers* (Markt Grafing, Bavaria: Buchdruckerei Hauser, 1945), 21.

61. *War Department Notes.*

62. Glassman, 21.

63. *Omaha Beachhead*, and Shae.

64. *Omaha Beachhead*, 106.

65. Parker.

66. *War Department Notes.*

67. Graves.

68. *War Department Notes.*

69. Ibid.

70. Ibid.

71. After Action Report, 5th Ranger Infantry Battalion. See copy, Appendix A.

72. Reville, John J., conversations.

73. *War Department Notes.*

74. Parker.

75. Sterne, Gary. Correspondence and Conversations.

76. Parker.

77. Sterne, conversations with John J. Reville.

78. Sterne.

79. Ibid.

80. After Action Report, 5th Ranger Infantry Battalion, 127 (Appendix A).

REFERENCES

M y sources and references are listed in the order in which they first appear in the story.

After Action Report, HMS *Prince Baudouin*. I obtained this report from Lieutenant Colonel Mark J. Reardon, who, upon learning of my interest in Ranger Operations on D-Day, gave them to me. There are seventeen reports in the series that include: HMS *Prince Charles*, HMS *Prince Leopold*, HMS *Prince Baudouin*, USS *PC-568*, USS *PC-567*, USS *Texas*, USS *Thompson*, and HMS *Tallybont*. The reports of the LSIs (Landing Ship Infantry) enclose subordinate reports such as Flotilla 507. A copy of the report of ML 304 is enclosed as Appendix G.

James W. Graves Jr. provided the needle that got me writing this story. He had been my communications platoon sergeant in the 5th Rangers. The sources marked "Graves" are from letters and discussions with Jim.

"Ace" Parker was one of the first contributors to this story. Author Jerry Astor was writing a book on the Invasion. Ace contributed an audio tape to Astor and later gave me a copy of the tape for use in my story. Later, when Ace wrote "Reflections of Courage on D-Day," I returned a copy of the tape to him. I used both the tape and "Reflections" as sources. Ace, a lieutenant, commanded A Company, 5th Ranger Infantry Battalion (5th RIB).

Robert W. Black, *Rangers in World War II* (New York: Ivy Books, 1992).

After Action Report of the 507th Assault Flotilla.

After Action Report of the USS *Texas.*

After Action Report of the 504th Assault Flotilla.

Hastings, Max, Sir. *Overlord: D-Day and the Battle for Normandy* (New York: Simon and Schuster, 1984), 80.

Omaha Beachhead (Washington, D.C.: Historical Division, War Department, Government Printing Office, 1945). OB is a splendid book, very authoritative, and quite contemporary with the actual events. The book covers all units engaged in the Omaha Beach landings, and part of my task was to ferret out those parts that referred to the 5th Rangers. Until I obtained *War Department Notes*, OB was my primary reference when I had no notes or memories of an event. The maps for this book are outstanding.

War Department Notes. Lieutenant Colonel Charles H. Taylor, a WD Historian, conducted interviews of the troops shortly after the Invasion. For the 5th Rangers, I was appointed to assist him. Taylor took copious notes during his interviews, and later, his assistant Sergeant Forrest Pogue organized and typed them into a usable form. These notes are the real source of facts about the Invasion. Kevan Elsby, an English author, formed a partnership with Joseph Balkoski, an American author, one doing archives research in the U.S., the other in the U.K. Joe Balkoski "discovered" these notes, and Elsby gave copies to me. The notes from Elsby covered only D-Day actions on Omaha Beach proper.

In 2005, Tom Hatfield, a University of Texas dean, contacted me in his research of a book on Earl Rudder. Tom gave me a copy of Taylor's notes for D+1 and the Point du Hoc operation! I used both sets of notes extensively.

Small Unit Actions (Washington, D.C.: Historical Division, War Department, 1946). One chapter in this pamphlet covers the actions at Pointe du Hoc in detail. It is superb if you want contemporary coverage of the 2nd Rangers fight at Pointe du Hoc.

Theodore Wells's letters to me. Wells's story, though in quotes, was rewritten to fit his story into the third person.

E-mail from Hicken.

Letters and conversations with Richard P. Merrill. Merrill was a captain in the 2nd RIB.

Bruce Bliven, Jr., *The Story of D-Day, June 6, 1944* (New York: Random House, 1994). One of the chapters covers the actions of C Company, 116th Infantry.

Conversations with Orman L. Kimbrough, Jr. Kimbrough's father was a lieutenant in Company C, 116th Infantry. When the 1st Battalion, 116th Infantry, was reorganized the night of D-Day, Lieutenant Kimbrough was given command of A Company, 116th Infantry.

Howard A. MacDonald, "Mine Was a Very Short War" (a memoir).

Conversations and e-mail with Thomas F. Herring. Herring was a mortar man in C Company, 5th RIB. He was also a longtime officer, secretary, and historian for the World War II Ranger Battalions Association.

Lieutenant Shea was aide-de-camp to General Cota during the Invasion. Shea wrote an extensive report on General Cota's actions on D-Day.

Ellias E. Reed, Jr., was a bangalore torpedo man in D Company, 5th RIB. Most of this vignette was from a telephone conversation with Bill.

Edward S. Luther commanded E Company, 5th RIB. Ed is the source of the story also found in WD Notes.

Henry S. Glassman was T/4 in Headquarters Company, 5th RIB. After the war he was assigned to write a history of the battalion. The result was *Lead the Way, Rangers*, a very useful coverage of the 5th's actions from activation to deactivation.

Sergeant Walter N. McIlwain was another early contributor to this story. "Mac" received a battlefield commission, stayed on in the army after the war and retired a lieutenant colonel. His story came from a letter to me.

Conversations with Stewart Bryant, a WWII historian.

S. L. A. Marshall, "First Wave at Omaha Beach," *Atlantic Monthly*, November 1960. This short story covers the landings of A and B Companies of the 116th Infantry.

Conversations with Victor H. Fast, a T/4 in Headquarters Company, 5th RIB.

Conversations with John J. Reville, a lieutenant in F Company, 5th RIB.

After Action Report, 5th RIB. Though it was signed by Hugo Hefflefinger, I actually wrote the report for the Normandy operation. A copy is at Appendix A.

Stern's interviews with John L. Burke, a staff sergeant in the medical detachment of Headquarters Company attached to A Company, 5th RIB.

Correspondence and conversations with Gary Sterne, an Englishman who has purchased most of the property on which the Maisy Battery was located. He has had extensive interviews with Rangers who fought at Maisy and with local Frenchmen who lived in the area at the time of the Invasion. He has also done extensive research into documents and maps pertaining to the Maisy area and the battle.

APPENDIX A

AFTER ACTION REPORT

5TH RANGERS BATTALION

———————

Jim Graves sent me the After Action Reports of the 5th Ranger Infantry Battalion from June 6, 1944, to March 31, 1945, as well as a copy of his list of units, their call signs, and frequencies. I remember writing the After Action Report for the D-Day operation and reproduce them here to show the official report on what happened to the battalion during the Normandy Campaign. In copying, I corrected a few obvious typographical errors.

HEADQUARTERS
FIFTH RANGER INFANTRY BATTALION
A. P. O. 655 U.S. ARMY
2 July 1944
SUBJECT: Action Against Enemy, Reports After/After Action
Reports.
TO: The Adjutant General, Washington, D.C.

1. The following is the story of the Fifth Ranger Infantry Battalion from the landing on the coast of France on D-Day, 6 June 1944, to 10 June 1944, which was the unit's last day of combat during the month of June.

D-DAY 6 JUNE 1944

At 0530 hours, the Fifth Ranger Infantry Battalion loaded into L.C.A.s from the mother ships, H.M.S. Prince Leopold and Prince Baudouin, and started the ten-mile run to the coast of France. The morale of the men was excellent, the weather cloudy, and the sea very choppy.

About five miles from shore, one (1) L.C.A. containing the First Platoon and a part of company headquarters of Company F, had shipped so much water that it was forced to drop out of the formation. This platoon did not make the assault landing with the battalion but did land near the St. Laurent-Sur-Mer Beach exit at 0900 after transferring to a passing L.C.T.

The beach was protected by a large number of under-water obstacles consisting of elements "C," hedgehogs and tetrahedra, many of which had Teller-Mines attached. Mortar and artillery shells were bursting in the area of these obstacles and a heavy concentration of small-arms fire swept the beach. A four-foot seawall ran laterally along the beach about 75 yards from the water's edge. Friendly troops were observed utilizing the protective cover afforded by this wall. A pall of smoke obscured the sharply rising ground immediately in the rear of and overlooking the beach. Our naval bombardment evidently had set fire to the vegetation covering the hill.

The L.C.A.s slowly threaded their way through gaps in the lines of obstacles and at H+75, 0745, the first wave consisting of one-half Battalion Headquarters, Companies A, B, and E, landed on Omaha Dog White Beach at a point approximately 800 yards East of Exit D-1. The Battalion Commander, Lieutenant Colonel Max F. Schneider, had ordered the flotilla commander to touch down his craft east of the intended landing point, Dog Green, because the tremendous volume of fire which covered that portion of the beach was inflicting a large number of casualties on the preceding wave.

The first wave crossed the beach in good order with few casualties, halted temporarily in rear of the seawall, and immediately reorganized.

The second wave, consisting of one half Battalion Headquarters, Companies C, D, and one (1) platoon of Company F, duplicated the feat of the first wave.

At a signal of the Battalion Commander the leading echelon scrambled over the wall, blew gaps in the protective wire, and protected from enemy observation by the curtain of rising smoke advanced unhesitatingly to a point near the top of the hill. Here the smoke had cleared and the topographical crest was being swept by effective automatic weapons fire. First Lieutenant Francis W. Dawson, Company D, led his platoon over the top and wiped out a strongpoint thereby enabling the battalion to advance.

Because of numerous minefields the battalion now changed-into a column formation and, after winding through their intricate pattern, the leading unit, Company B, reached the St. Laurent-Sur-Mer—Vierville-Sur-Mer road at a point approximately one (1) kilometer east of Vierville-Sur-Mer. During the advance numerous Germans, well concealed in weapons pits constructed in hedgerows, were killed.

Company B advanced toward Vierville-Sur-Mer receiving heavy sniper and machine gun fire. Several direct hits from enemy artillery on the rear of the battalion column caused numerous casualties. Company E attempted a penetration to the south but was halted by intense machine gun fire. An 81-mm mortar concentration fired by Company C knocked out several of these positions but they were rapidly replaced and the advance remained halted.

The weight of the attack was shifted toward Vierville-Sur-Mer and, after overcoming considerable sniper resistance, the battalion advanced through the village to its western outskirts where it was again held up by a large volume of concerted machine gun and sniper fire. At this point contact was established with the Commanding Officer First Battalion 116th Infantry and approximately 150 men of his unit. Dusk was falling and the battalion was ordered to dig-in a perimeter defense for the night. Companies A, B, and C of the Second Ranger Battalion, numbering approximately 80 men, also assumed a portion of the defense area. Tanks of the 743rd Tank Battalion moved within the defense area. Except for occasional exchanges with enemy snipers and machine guns the night was one of little activity.

One (1) platoon of Company A which became separated from the battalion after crossing the seawall proceeded through Vierville-Sur-Mer to the rallying point southwest of town, arriving there at 1600 hours. In accomplishing this feat they captured 12 Germans and killed at least an equal number. Leaving the rallying point shortly thereafter this unit fought its way through to Pointe du Hoc (the Battalion objective) and contacted the Second Ranger Battalion, arriving there about 2200 hours.

The platoon from F Company which landed near St. Laurent-Sur-Mer received a large amount of artillery and machine gun fire on the beach. Patrols were sent out to locate the remainder of the Fifth Battalion but were unable to gain contact. Attempting to move along the beach toward Vierville-Sur-Mer this platoon was subjected to artillery fire, receiving eight casualties. After advancing 600 yards to the west the unit was engaged by a superior force and pinned down. When darkness fell the platoon retained this position.

The results for the first day were about 100 prisoners taken, 150 enemy dead, and approximately 60 Rangers killed and wounded.

D+1—7 JUNE 1944

Plans were made for enlarging the beach-head and for relieving the three companies of the Second Ranger Battalion at Pointe du Hoc.

At 0600 a force composed of 80 men from the Second Ranger Battalion (Companies A, B, and C), Companies C and D of our battalion, 150 men from the First Battalion 116th Infantry, and six tanks of the 743rd Tank Battalion, advanced toward Pointe du Hoc. Encountering harassing sniper fire, this force advanced to a point approximately one (1) kilometer west of St. Pierre Du Mont where it received a large concentration of artillery fire. This fire continued falling from 1000 hours until 1800. The force withdrew to St. Pierre Du Mont and set up a defense in the town. Under cover of darkness a two-man patrol moved through enemy lines to Pointe du Hoc, contacted the Commanding Officer Second Ranger Battalion, and returned laying wire to establish communication between forces.

The remainder of the battalion was given the mission of improving the beach-head.

Company B resumed the attack to the southwest from the western edge of Vierville-Sur-Mer at 0630. This unit was not able to advance far but did knock out several machine gun nests and numerous snipers.

The remaining platoon of Company A and the remaining platoon of Company F supported by four tanks of the 743rd Tank Battalion attacked South from the town wiping out snipers, machine gun nests, and several enemy combat patrols. Approximately 25 of the enemy were killed and 85 were captured.

During the night, snipers had infiltrated back into the town so Company E cleaned out the town again.

At approximately 1100 hours these units were ordered to set up a defense around the town. At 1400 hours, Company E, which was defending the eastern portion of the town beat off a determined counterattack of about company strength.

At 1900 hours elements of the 116th Infantry moved into the eastern half of the sector and the units of this battalion shifted to the right to defend only the remaining half of the sector.

The A Company Platoon on Pointe du Hoc assisted in repulsing three counter-attacks early this morning. A seven-man patrol from this platoon made an unsuccessful attempt to infiltrate through the enemy positions to contact the Fifth Ranger Battalion. The remainder of the day was spent manning a portion of the defense area.

The first platoon of Company F attacked inland from the beach at 0800 and by 1400 hours secured their objective, destroying three pillboxes and several weapons emplacements. Eight Germans were killed and 36 captured. At this time the platoon was contacted by Major Street of Admiral Hall's staff, loaded into a L.C.V.P. with food, water, and ammunition, and transported to Pointe du Hoc contacting the Commanding Officer Second Ranger Battalion at 1700 hours. An eight-man patrol from this platoon infiltrated through the enemy positions and by 0800 of D+2 had contacted the force at St. Pierre-du-Mont.

Results for the second day were approximately 150 prisoners taken, 80 killed, and 40 Ranger casualties.

D+2—8 JUNE 1944

At 0100 hours orders were received to prepare to move to Pointe du Hoc at 0600. At 0630 hours a force composed of two battalions of the 116th Infantry and three companies of the Fifth Ranger Battalion advanced from Vierville-Sur-Mer down the road west toward Pointe du Hoc. No resistance was encountered and the force at St. Pierre-du-Mont was contacted at 0815.

Companies B and E were given the mission of taking and

holding the high ground west of the Sluice Gate at Grandcamp-les-Bains. These companies in column, Company B leading, advanced through the low ground south of the East-West road leading into Grandcamp at 1000. Initially no fire was received and the town appeared to have been deserted by the enemy. The leading elements of the two companies approached to within 25 yards of the bridge where the force was pinned down by a heavy concentration of mortar and machine gun fire. The two companies withdrew to positions on the high ground east of the Sluice Gate Bridge where they were joined by Company D, which had just returned from Pointe du Hoc. They were passed through by the Second and Third Battalions of the 116th Infantry, which supported by tanks, artillery fire, and naval gunfire, successfully captured the town of Grandcamp-les-Bains. D Company and E Company went into defensive positions protecting the Sluice Gate Bridges and mopped up positions along the coast toward Pointe du Hoc. B Company occupied a portion of the all around defense set up by the Second Ranger Battalion on the high ground east of the Sluice Gate Bridge.

At 0900 Companies C, D, one platoon of F Company, the remainder of A, B, C, of the Second Ranger Battalion, and one platoon from Company A advanced toward Pointe du Hoc to assist the three companies of the Second Ranger Battalion. Meeting no resistance they contacted that unit. Companies A and F were now complete organizations.

Shortly thereafter the Rangers were brought under fire by the Third Battalion 116th Infantry and tanks of the 743rd Tank Battalion. This force was attacking Pointe du Hoc from the southwest and inflicted six casualties on our force, two of which were killed.

D Company advanced toward Grandcamp-les-Bains and joined Companies B and E.

Companies A, C, and F advanced east and south of the inundated area toward Maisy as part of a force consisting of the First Battalion 116th Infantry. Two half-tracks of the Second Ranger Battalion were attached to the Rangers. This force halted one-half mile northeast of Maisy for the night, meeting no resistance en route.

Results for the third day were approximately 20 Germans killed and 35 captured. The battalion had 10 casualties.

D+3—9 JUNE 1944

Company E continued to mop up positions along the coast toward Pointe du Hoc, killing several Germans and capturing about 40.

At 1300 hours Companies B, D, and E marched directly south from Grandcamp on a secondary road, changed direction to the west at Le Manoir, again changed direction to the south on the Isigny road, and went into bivouac about 400 yards west of Osmanville.

A, C, and F Companies, detached from the First Battalion 116th Infantry which had by-passed the battery position southwest of Maisy, were given the mission of cleaning out that strong point. They were supported on this mission by the two half-tracks Second Ranger Battalion, Company B 81st Chemical Weapons Battalion, and four 81-mm mortars carried by Company C. A concentration by the 58th Field Artillery Battalion preceded the attack. Attacking with two companies in column the position was successfully carried. The strong-point contained three 105-mm howitzers, numerous small arms, large stocks of ammunition and food, and ejected approximately 90 prisoners. Shortly thereafter this force marched to the bivouac area west of Osmanville where it joined the remainder of the battalion at 2000 hours.

At 2100 a three-man patrol from Company E was dispatched to reconnoiter the Light Engineer Bridge across the Vire River. This patrol accomplished its mission and returned at 2400 hours.

Results for the fourth day were approximately 20 Germans killed, 130 prisoners, and 18 Ranger casualties.

D+4—10 JUNE 1944

At approximately 0430 hours the battalion bivouac area was bombed and the unit suffered three casualties.

Companies C, D, and F were given the mission of mopping up the coastal fortifications from Grandcamp-les-Bains to Isigny. They moved out at 0830 and meeting little resistance, returned at 1530 with approximately 200 prisoners.

Patrols in the vicinity of the battalion area captured approximately 35 Germans. Results for the fifth day were 235 Germans captured and 6 Ranger casualties.

2. In five days of fighting this battalion had 23 men killed, 89 wounded, and 2 missing. Approximately 850 prisoners were taken and 350 Germans killed.

For the Commanding Officer:

Hugo W. Heffelfinger
Capt. Infantry

THE STORY OF FATHER LACY

The 5th Rangers had been looking for a chaplain for some time, and now it was clearly too late, D-Day had to be soon. We were already locked up in our base camps. The officers and non-coms had already seen the maps and sand tables of the invasion beaches. True, the names of all the places had been expunged, but who could miss a shoreline running east and west, no major towns or highways. A rural area . . . not the Pas de Calais, it had to be Normandy.

Then suddenly, he was there. Chaplain (First Lieutenant) Joseph R. Lacy. "Red! Come over here and meet our new chaplain," called Major Sullivan, the battalion executive officer. I looked at Lacy. Old, probably in his late thirties or early forties. Short, he couldn't have been over five foot six. Fat, at least thirty pounds overweight. Thick glasses. Dripping with perspiration. This was our Ranger chaplain!

"He's yours." And so I took him with me to find a bunk. Turned out he was a Roman Catholic, a good thing in the 5th Rangers because the majority of our officers and about half of the enlisted men had come to us from the YD, the Yankee Division, and most of them were Catholics. I found a bunk for him and helped him unpack, noted what equipment he lacked, and sent someone to draw his requirements.

Our conversation was short and to the point. "Padre, you aren't in very good shape."

"No, I'm not."

"Do you know what our mission is?"

"No."

"Well, we don't either," I continued, "but we have a pretty good idea that it involves a long, fast march across country, through enemy territory. Do you think you can keep up with us?"

"Probably not, but if you leave enough signs, I'll catch up to you when you stop." And that was it.

I went back to Sullivan and said, "Sully, we really got ourselves a hot one there. I just don't think he can hack it."

"You're probably right, Red, but at least we have a chaplain for these last few days before the invasion."

And so we did. Lacy was a really enjoyable guy, his only problem was his physical condition. He'd never keep up with us. "Padre, you know you can't keep up with us, and you know we can't wait for you."

"True, but don't worry about me, I'll get there sooner or later."

And so it was. Lacy kept out of the way in those last hectic days before we boarded the HMS *Prince Baudouin*. He established a rapport with the men that was much needed. Protestant, Catholic, Jewish, it didn't matter to him or to the men, they loved him and his unassuming sympathy and empathy.

Aboard the *Baudouin*, I didn't see much of him. During the day, we—officers and men—were busy standing watch, maintaining our equipment, checking the men's, studying the terrain we were about to experience. By now the maps had names on them. In the evening, while the officers congregated in the ward room with the British officers, drinking Scotch whisky and singing mostly Irish songs, the Padre was down among the men giving solace, comfort, advice. He was assigned to my boat, so I checked him and his equipment out a dozen times as we went through our boat drills.

And then, "Attention on deck! Attention on deck! United States Rangers, embarkation stations! United States Rangers, embarkation stations!" Father Lacy loaded with us, but other than making sure he was aboard, I paid no attention to the little old fat man.

The next time I saw him, I was kneeling on Omaha Dog Red Beach right next to the seawall, looking back at my LCA as my men still poured out of it and began running toward me and the safety of that wall. There was Father Lacy, the last man out. He was no more than ten feet clear of the boat when a German shell hit the fantail of the LCA. The Padre was unhit, but the British crew must have been killed. I looked away and did not see Father Lacy again until much later.

Others saw him and like minstrels sang his praises. Lacy didn't cross the beach like we heroes did. He stayed down there on the water's edge where the artillery was falling, pulling the wounded forward, ahead of the advancing tide. He comforted the dying.

He calmly said prayers for the dead. He led terrified soldiers to relative safety behind debris and wreckage, half carrying them, half dragging them, binding up their wounds. Never once did he think of his own safety, always helping those that needed his help to survive that awful inferno.

The 5th Rangers left the beach a few minutes later, but Father Lacy stayed behind at the water's edge, doing the work for which God had chosen him.

True to his word the Padre caught up with us later. He was delayed, he said.

Chaplain Lacy was awarded the Army Distinguished Service Cross for extraordinary heroism in action. The citation is contained in Appendix D.

APPENDIX C

HEADQUARTERS ROSTER

—————

This appendix contains four rosters, which together list all the members of Headquarters Company, 5th Ranger Infantry Battalion who participated in the Invasion.

The first roster is that of all personnel who made the assault landing on D-Day, landing between 0745 and 0750 on Omaha Dog White and Dog Green beaches. About half these men were in Lieutenant Colonel Schneider's LCA from the HMS *Prince Leopold* and half were from Major Sullivan's LCA 1377 from the HMS *Prince Baudouin*. Most of those in the "Detached to Group" category landed with the 2nd Rangers in the first wave of Ranger Force C at 0735. At least one, Wells, landed at Pointe du Hoc at 0710.

The second list contains those Headquarters Rangers who landed with the battalion vehicles at approximately midnight of D-Day.

The third list contains the Battalion Rear Echelon, which landed after D-Day. There were twenty-nine enlisted men in the rear echelon, mostly cooks and truck drivers. However, the list shows thirty-six names. At most, seven of these men joined Headquarters Company after the Invasion.

The lists are derived from five rosters still in my possession:

a. *A handwritten list. Except for handwriting errors, it is accurate. This list contains only surnames.*

b. *A Soldier Voting Roster dated August 16, 1944.*

c. *An undated list written before Group Headquarters was dissolved.*

d. *Glassman's,* Lead the Way, Rangers.

e. *Black's,* Rangers in World War II.

The handwritten list is the one that I carried in the Invasion. It contains all the names of the men in Headquarters except for the attached JASCO (293rd Joint Assault Signal Company) Team and the company runners assigned to Headquarters. The JASCO Team was attached to us to provide forward observers and fire control for naval gunfire. Whenever I received a report about a casualty, I recorded it on this sheet.

The Tables of Organization (TOE 7-85, dated February 25, 1944) for the 5th Rangers had six companies of three officers and sixty-five enlisted men. Headquarters and Headquarters Company had eight officers and eighty-eight enlisted men. An attached Medical Detachment had one officer (the battalion surgeon) and eleven enlisted men. This brought the Headquarters and Headquarters Company to a grand total of 108 officers and men.

ROSTER OF HEADQUARTERS, 5TH RANGER INFANTRY BATTALION

June 6, 1944

1. Roster of Rangers in the Assault Waves

Rank	Name
HEADQUARTERS	(5-O)
Lt Col	Schneider, Max F.
Capt	Butler, Edmund J.
Capt	Byrne, William P., Jr.
Capt	Heffelfinger, Hugo W.
Lt	Askin, Stanley L.
DETACHED TO GROUP	(3-O, 11-E)
Maj	Sullivan, Richard P. WIA
1 Lt	Lacy, Joseph R. (Chaplain)
Lt	Swazey, Vaughn R.
SSgt	Brewer, UNK
SSgt	Button, LeRoy T. WIA-Evac
T/4	Tobin, John J., Jr.
T/5	Colbath, John E.
T/5	Schapp, UNK
	(Dan D, Schopp later in F Co)
T/5	Walters, Ted M. WIA-Evac
T/5	Wells, Theodore H. WIA-Evac-Dy
T/5	Rohlin, David W.
Pfc	Cunnally, Robert J.
Pfc	Wentworth, LeRoy L.
Pfc	Piekarz, Henry P.
HQ COMPANY	(2-O, 29-E)
Capt	Raaen, John C., Jr.
Lt	Van Riper, Howard E.
MSgt	Dean, Minor C.
MSgt	Dunkle, Harry J.

TSgt	Epstein, Herbert NMI
TSgt	McGuire, Wilfred F.
TSgt	Woodill, Russell NMI
SSgt	Moore, Alfred E.
Sgt	Graves, James W., Jr.
T/4	Brown, Lee M.
T/4	Fast, Victor H.
T/4	Fitzgerald, James V.
T/4	Jenner, William H.
T/4	Whiteley, Fred O.
Cpl	Soper, James F.
T/5	Bowser, Macey NMI
T/5	Curley, William P.
T/5	Halacy, Thomas E., Jr.
T/5	James, William H.
T/5	Kostowski, Edward P.
T/5	Krumenacker, Edward C. WIA-Evac
T/5	Munley, John V.
T/5	Pyles, Charles R.
Pfc	Davis, Everett B., Jr.
Pfc	Derone, Joseph WIA
Pfc	Heidekrueger, Arthur W.
Pfc	Hirth, Charles NMI
Cpl	Sharp, Jack L.
Pvt	Gwiadowski, Leonard NMI
Pfc	Coughlin, Francis WIA-Evac
Pfc	McCullough WIA-Evac

ATTACHED MEDICS	(1-O, 3-E)
Capt	Petrick, Thomas G.
SSgt	Bartlett, Frank R., Jr.
T/4	Mullin, Peter V.
T/4	Clawson, David L.

6 RUNNERS FROM COMPANIES ATTACHED to HQ   (6-E)
Names unknown.

Names unknown.

2. Roster of Rangers in the Mid-Night Follow-on Wave.

S-4 Section and MOTOR POOL (HQ Co)   (2-O, 35-E)

Capt	Murray, William E.
Lt	Nee, Richard J.
TSgt	Zack, Walter T. KIA
Pfc	Caraber, Andrew J. WIA-Evac
SSgt	Jones, Carl NMI
Sgt	Knutson, Emmett E.
Sgt	Powell, Francis D.
T/4	Graves, Raymond C.
T/4	Hiffner, Keith E.
Cpl	Lewis, Harold A. WIA-Evac
T/4	Seeley, Norman C.
T/5	Westhoff, John K.
T/5	Barrows, Richard H.
T/5	Barry, Gerald T.
T/5	Dowd, Theodore NMI
Pvt	Ingram, Jesse L.
T/5	Kane, William L.
Pfc	Lanham, Thomas E.
T/5	Madore, Lawrence R.
T/5	Nutkins, William E.
T/5	Trahan, Wilfred J.
Pfc	Barton, Harold E.
Pfc	Douglas, Paul C.
Pfc	Dunham, Harry R. WIA-Evac
Pfc	Dwyer, Thomas M.
Pfc	Ekern, Howard D.
Pfc	Harwood, Robert E.
Pfc	Moyer, Harry L.
Pfc	Robbins, Charles L.
Pfc	Royle, James E.
Pfc	Schultz, John N.

T/5	Peseroff, Roy I.
Pfc	Wilson, Warren H.
Pvt	Hacker, UNK WIA
Pvt	Ritchie, Balis L.
Pvt	Wiseley, Dean V. WIA
Pvt	Woods, Wayne P.

MEDIC ATTACHED TO MOTOR POOL   (1-E)

T/5	Germain, Irvin L.

3. Rangers in the Rear Echelon

REAR ECHELON	(29-E)
TSgt	Livingston, Herbert D.
TSgt	Hall, Harry C.
SSgt	Hathaway, Richard H., Jr.
T/4	Bazley, Lawrence F.
T/4	Bridges, George P., Jr.
T/4	Lewis, Raymond NMI
T/4	Loftin, Leon H.
T/4	Scarbrough, John J.
Cpl	Carroll, Edmund L.
Cpl	Keefe, George L.
Cpl	Siatkowski, Francis A.
T/5	Ballard, Eula C.
T/5	Bartling, Howard J.
T/5	Hedrick, Richard H.
T/5	Landers, Lonnie H.
T/5	Long, Jack E.
T/5	Manning, George E.
T/5	Roller, Nathan H.
T/5	Stone, Roger A.
T/5	Weatherford, Richard M.
Pfc	Akin, James H., Jr.
Pfc	Biava, Louis A.
Pfc	Frazier, Joe R.
Pfc	Hasselback, Harold L.

Pfc	Jarrell, Edward R.
Pfc	Nelson, Donald T.
Pfc	Paradis, Alouise A.
Pfc	Simonette, Nicholas J.
Pvt	Crim, Harold W.
Pvt	King, Harold L.
Pvt	Kramlich, Samuel E., Jr.
Pvt	Nenna, John M.
Pvt	Talley, Clyde W.
Pvt	Templeton, Calvin C.
Pvt	Westfall, George R.
Pvt	Zaher, Stephen NMI

4. Other Headquarters Rangers

DETACHED SERVICE	(1-E)
Cpl	Butterfield, Victor J., Jr.

LISTED UNKNOWN	(1-E)
Pfc	Malloy, Paul F. (later found in E Co) WIA

APPENDIX D

AWARDS

CITATIONS FOR THE DISTINGUISHED SERVICE CROSS

This list contains the names of the members of the 5th Ranger Infantry Battalion who were awarded the Distinguished Service Cross for "extraordinary heroism in action" during the D-Day Campaign. The Individual Citations are on the following pages.

Lt. Col. F. Max Schneider	Battalion Commander
Maj. Richard P. Sullivan	Battalion Executive Officer and S-3
Capt. George P. Whittington	B Company Commander
1st Lieut. Charles H. Parker	A Company Commander
1st Lieut. Joseph R. Lacy	Chaplain
1st Lieut. Francis W. Dawson	Platoon Leader, D Company
Sgt. Denzil O. Johnson	A Company
Sgt. Willie W. Moody	C Company
T/5 Howard D. McKissick	C Company
Pfc. Alexander W. Barber	Medical Detachment

HEADQUARTERS
FIRST UNITED STATES ARMY
APO 230
GENERAL ORDERS (No. 28) 20 June 1944

SECTION
Award of Distinguished-Service Cross I
Amendment of General Orders   II

I—AWARD OF DISTINGUISHED-SERVICE CROSS
Under the provisions of AR 600-45, 22 September 1943, and
pursuant to authority contained in paragraph 3c, Section I,
Circular No. 32, Hq ETOUSA, 20 March 1944, as amended, the
Distinguished-Service Cross is awarded to the following officers
and enlisted men

EXTRACT

Lieutenant Colonel Max F. Schneider, O384849, Infantry,
[5th Ranger Infantry Battalion], United States Army, for
extraordinary heroism in action on 6 June 1944 at [Omaha Dog
Red Beach], France. In the initial landings in the invasion of
France, Lieutenant Colonel Schneider led the 5th Ranger Infantry
Battalion ashore at "H" Hour on "D" Day in the face of extremely
heavy enemy rifle, machine gun, mortar artillery and rocket
fire. Upon reaching the beach Lieutenant Colonel Schneider
reorganized his unit. During this reorganization he repeatedly
exposed himself to enemy fire. He then led his battalion in the
assault on the enemy beach positions, and having accomplished

this mission led them up a steep incline to assault the enemy gun emplacements on top of the hill. The destruction of these enemy positions opened one of the vital beach exits, thereby permitting the troops and equipment which had been pinned down to move inland from the beach, with the result that reinforcements could be landed from the sea. By his heroic leadership and personal courage Lieutenant Colonel Schneider set an inspiring example to his command, reflecting the highest traditions of the armed forces. Entered military service from Iowa.

EXTRACT

Major Richard P. Sullivan, O399856, Infantry, [5th Ranger Infantry Battalion], United States Army, for extraordinary heroism in action from 6 June 1944 to 10 June 1944 near [Vierville-sur-Mer and Maisy], France. Completely disregarding his own safety, he personally directed a successful landing operation and lead [sic] his men across the beach covered with machine gun, artillery and rocket fire. After reorganizing his men he immediately resumed his duties as Battalion Executive officer and was placed in command of two Ranger companies which fought their way inland against fierce opposition to join and relieve the Ranger detachment on [Pointe du Hoc]. After laying communications through enemy lines under cover of darkness, Major Sullivan directed the Rangers' progress across country to [Pointe du Hoc] and [Grandcamps-les-Bains]. In cooperation with United States Infantry an attack was begun on the [Maisy] battery. When certain elements were temporarily halted by artillery fire Major Sullivan, who had been wounded at [Omaha Beach], calmly and courageously rallied his officers and men, ordered a renewal of the attack, and instead of bypassing the resistance, advanced over heavily mined terrain to capture the [Maisy] battery with a loss of only fifteen (15) men. Eighty-six (86) prisoners and several large caliber artillery pieces in concrete bunkers were taken. Attacks by Major Sullivan's command contributed greatly to the success of the entire Corps' operations. By his intrepid direction, heroic leadership and superior professional ability, Major Sullivan set an inspiring example to his command reflecting the highest traditions of the armed forces. Entered military service from Massachusetts.

EXTRACT

Captain George P. Whittington, O403921, Cavalry, [5th Ranger Infantry Battalion], United States Army, for extraordinary heroism in action on 6 June 1944 at [Omaha Dog Red Beach], France. Captain Whittington commanded a Ranger company [Company B, 5th Ranger Infantry Battalion] which landed on the coast of France at "H" hour. The landing was made on the beach against heavy rifle, machine gun, mortar, artillery and rocket fire of the enemy. Despite this fire, he personally supervised the breaching of hostile barbed wire and obstacles by the use of bangalores. He then led his company and the remainder of his battalion through the gap created. He then directed the scaling of a 100-foot cliff by his company. When he reached the top of the cliff he crawled under enemy machine gun fire and destroyed the enemy position. Captain Whittington's bravery, aggressiveness and inspired leadership are in keeping with the highest traditions of the service. Entered military service from Arkansas.

EXTRACT

First Lieutenant <u>Francis W. Dawson</u>, 0400036, Infantry, [5th Ranger Infantry Battalion], United States Army, for extraordinary heroism in action on 6 June 1944, at [Omaha Beach], France. Lieutenant <u>Dawson</u> led his Ranger platoon ashore in the invasion of France against heavy enemy artillery, machine gun, and small-arms fire. He then personally took charge of the breaching of wire entanglements. When a gap was created, he led his platoon through it and directed them in scaling a 100-foot cliff. Upon reaching the top of the cliff, he, accompanied by one soldier, rushed forward with a submachine gun and destroyed a German pill box, killing or capturing the enemy located therein. Lieutenant <u>Dawson's</u> heroic action and aggressive leadership are in keeping with the highest traditions of the service. Entered military service from South Carolina.

EXTRACT

First Lieutenant <u>Joseph R. Lacy</u>, O525094, Chaplain Corps, [5th Ranger Infantry Battalion], United States Army, for extraordinary heroism in action on 6 June 1944 at [Omaha Dog Beach], France. In the invasion of France, Chaplain <u>Lacy</u> landed on the beach with one of the leading assault units [the 5th Ranger Infantry Battalion]. Numerous casualties had been inflicted by the heavy rifle, mortar, artillery and rocket fire of the enemy. With complete disregard for his own safety, he moved about the beach, continually exposed to enemy fire, and assisted wounded men from the water's edge to the comparative safety of a nearby seawall, and at the same time inspired the men to a similar disregard for the enemy fire. Chaplain <u>Lacy's</u> heroic and dauntless action is in keeping with the highest traditions of the service. Entered military service from Connecticut.

EXTRACT

First Lieutenant <u>Charles H. Parker</u>, 01290298, Infantry, [5th Ranger Infantry Battalion], United States Army, for extraordinary heroism in action on 6, 7, and 8 June 1944 from [Vierville-sur-Mer to Le Pointe du Hoc], France. In the invasion of France, Lieutenant <u>Parker</u> led his company up the beach against heavy enemy rifle, machine gun and artillery fire. Once past the beach he reorganized and continued inland. During this advance numerous groups of enemy resistance were encountered. Through his personal bravery and sound leadership this resistance was overcome, and his company succeeded in capturing [Le Pointe du Hoc], the Battalion objective. The following morning, Lieutenant <u>Parker</u> led a patrol through enemy held territory in an effort to establish contact with the balance of the Battalion. Lieutenant <u>Parker's</u> valor and superior leadership are in keeping with the highest traditions of the service. Entered military service from South Dakota.

EXTRACT

Sergeant <u>Denzil O. Johnson</u>, 38452897, Infantry, [5th Ranger Infantry Battalion], United States Army, for extraordinary heroism in action on 7 June 1944 at [Pointe du Hoc], France. Sergeant <u>Johnson</u> was scout for a patrol sent out to bring reinforcements to the isolated remnants of a Ranger battalion. The patrol was driven to the edge of a high cliff along the sea by enemy machine gun fire. Led by Sergeant <u>Johnson</u> the patrol worked themselves along the face of the cliff. When the patrol came to the top of the cliff it was stopped by hostile machine gun fire and a minefield. With complete disregard for his own personal safety Sergeant <u>Johnson</u> across the fire swept ground seeking a path through the minefield. Finding an escape route Sergeant <u>Johnson</u> returned to the patrol and led it to a position of safety. Sergeant <u>Johnson's</u> heroic and valorous action is in keeping with the highest traditions of the service. Entered military service from Arkansas.

EXTRACT

Sergeant <u>Willie W. Moody</u>, 93628019, Infantry, [5th Ranger Infantry Battalion], United States Army, for extraordinary heroism in action on 7 June 1944 and 8 June 1944 in France [at Pointe du Hoc]. Sergeant <u>Moody</u> volunteered to attempt to make contact with a battalion of Rangers that had been cut off. At midnight Sergeant <u>Moody</u> moved off and started through enemy lines. After several hours of crawling through enemy minefields, enemy outposts and enemy installations, he finally contacted the battalion that had been cut off. He then returned to his own unit, and started out again with a reel of wire to a position where he believed his mortars would be set up. He returned through the enemy lines unreeling the wire so that accurate fire could be placed upon the enemy positions once the mortars had been placed in position. Sergeant Moody's heroic action is in keeping with the highest traditions of the service. Entered military service from Virginia.

EXTRACT

Private First Class Alexander W. Barber, 33575048, Medical Corps, [Medical Detachment, 5th Ranger Infantry Battalion], United States Army, for extraordinary heroism in action on 6 June 1944 in France [Omaha Dog Beach]. Private First Class Barber landed with his medical unit on the coast of France at a time when the beach was under heavy enemy rifle, machine gun and artillery fire. Numerous casualties had already been inflicted by this devastating fire. In spite of this heavy fire, Private First Class Barber constantly exposed himself to the direct fire of the enemy as he went along the beach administering aid to the wounded. On one occasion he took a horse and cart into the middle of an artillery barrage to bring out three (3) men who had been wounded. Private First Class Barber's heroic and gallant action is in keeping with the highest traditions of the service. Entered military service from Pennsylvania.

EXTRACT

Technician Fifth Grade Howard D. McKissick 39204430, Fifth Ranger Infantry Battalion, United States Army. For extraordinary heroism in action against the enemy on 7 June 1944, in France. Technician Fifth Grade McKissick volunteered with another man to make contact with a battalion of Rangers that had been cut off. At midnight Technician Fifth Grade McKissick moved off and started through enemy lines. After several hours of crawling through enemy minefields, enemy outposts, and enemy installations he finally reached the isolated battalion. He and his comrade then returned to their own unit, and started out again with a reel of wire to a position where they believed their mortars would be set up. He and his companion returned through the enemy lines unreeling the wire so that accurate fire could be placed on the enemy positions once the mortars were set up. Technician Fifth Grade McKissick's heroic action reflects credit on himself and is in keeping with the highest traditions of the Armed Forces. Entered military service from Washington.

By command of the ARMY COMMANDER W. B. KEAN, Major General, G.S.C., Chief of Staff
OFFICIAL:

/s/
R. S. Nourse,
Colonel, AGD,
Adjutant General.

Editor's note: Locations in brackets were considered classified and omitted from the original order. These locations have been supplied by the editor, and may not be correct. The unit designations in brackets were added by the editor. The author wrote the original citations of all but Lieutenant Colonel Schneider and Major Sullivan.

APPENDIX E

BATTLE HONORS

GENERAL ORDERS   (WAR DEPARTMENT, No. 73)
Washington, D.C., 6 September 1944.

EXTRACT

II-BATTLE HONORS.—l. As authorized by Executive Order No. 9396 (Sec. I, Bull. 22, WD, 1943), superseding Executive Order No. 9075 (Sec. III, Bull. 11, WD, 1942), citation of the following unit in General Orders, No. 36, Headquarters 1st Infantry Division, 13 July 1944, as approved by the Commanding General, First Army, is confirmed under the provisions of section IV, Circular No. 333, War Department, 1943, in the name of the president of the United States as public evidence of deserved honor and distinction. The citation reads as follows:

The 5th Ranger Infantry Battalion is cited for outstanding performance of duty in action. In the invasion of France the <u>5th Ranger Infantry Battalion</u> was assigned the mission of securing a sector of the beachhead. As the landing assault unit in this sector the battalion landed on the beach at H—hour on D—day.

This landing was accomplished in the face of tremendous enemy rifle, machine gun, artillery, and rocket fire. In addition, the battalion encountered mines and underwater and beach obstacles. Refusing to be deterred from its mission of securing a beachhead, the <u>5th Ranger Infantry Battalion</u> faced concentrated enemy fire and hazardous beach obstacles with determination and gallantry.

Although subjected to heavy enemy fire during the entire day and despite numerous casualties and fatigue, the courage and esprit de corps of this battalion carried the enemy positions by nightfall, thereby securing the necessary beachhead without which the invasion of the continent could not proceed. The heroic and gallant action of the <u>5th Ranger Infantry Battalion</u> in accomplishing this mission under unusual and hazardous conditions is in keeping with the highest traditions of the service.

BY ORDER OF THE SECRETARY OF WAR:
G. C. MARSHALL, Chief of Staff.

OFFICIAL:
ROBERT H. DUNLOP,
Brigadier General,
Acting The Adjutant General.

APPENDIX F

CITATION FOR THE SILVER STAR

CAPTAIN JOHN C. RAAEN, JR., 025486, Corps of Engineers, Headquarters Company, Fifth Ranger Infantry Battalion, United States Army. For gallantry in action in the allied invasion of France, 6 June 1944. As Headquarters Commandant of his Battalion, a part of the initial assault wave, Captain Raaen landed on the beach near Vierville-sur-Mer, France, under a ruinous and devastating barrage of observed enemy machine gun, sniper, mortar and artillery fire from well concealed and strategically located positions on the inner beach and on the cliffs overlooking the beach. Acting in the midst of reigning confusion, he repeatedly exposed himself without the least regard for his personal safety, to the full view of the enemy as he calmly moved along the beach reorganizing men who were without officer control into small groups which he urged forward to the assault. Having gone some distance inland with his Battalion, Captain Raaen, on one occasion, worked his way back across the desolated area to the beach under intense sniper fire, securing a vehicle which he loaded with badly needed ammunition and driving it to his troops who were attacking St. Pierre du Mont. The inspiring courage displayed by Captain Raaen on the beach, and his heroic achievement in securing ammunition under highly adverse conditions are examples of gallantry which reflect the highest credit upon him and upon the Military Service. Entered Military Service from: Arlington, Virginia.

APPENDIX G
EXTRACT OF ORDERS

HQ, RCT 116
APO #29, USARMY
11 May 1944

TOP SECRET
"NEPTUNE"
3 of 6 Pages

<u>d</u>. RANGER Group.

(1) Force "A," consisting of 3 companies (reinf), at H-Hour D-Day land on Beach CHARLEY, capture POINTE DU HOE [sic], prepared to repel counter-attack, and cover advance of remainder of Ranger Group. Par <u>d</u> (3) below.

(2) Force "B" consisting of 1 Company land at H+3 minutes D-Day on Beach OMAHA DOG GREEN, move rapidly through breach in wire on western edge of DOG GREEN BEACH, destroy defenses at POINTE ET RAZ DE LA PERCEE. Continue advance along coast, assist lst Bn 116th Inf in destruction of fortifications from POINTE ET RAZ DE LA PERCEE to and including POINTE DU HOE. On arrival at POINTE DU HOE revert to Ranger Group.

(3) Ranger Group less detachments (Force "C") above, land at H+60 D-Day and execute one of following plans:

PLAN 1. If success signal is received prior to H+30 from FORCE "A" par d (1) above, FORCE "C" will move to and land at H+60 minutes D-Day on BEACH CHARLEY. Gain contact with FORCE "A" destroy fortifications at POINTE DU HOE and along coast from POINTE DU HOE to SLUICE GATE (563933). Reorganize, prepare to repel counter-attack. Cover advance of the lst Bn 116th Inf to POINTE DU HOE. Upon arrival of lst Bn 116th Infantry at POINTE DU HOE, Ranger Group is attached to lst Bn 116th Infantry.

PLAN 2. Force "C" land on BEACH OMAHA DOG GREEN commencing at H+60 minutes D-Day move rapidly to POINTE DU HOE. Capture POINTE DU HOE. Destroy fortifications from POINTE DU HOE to SLUICE GATE (563937-563933) reorganize, prepare to repel counter-attack. Cover advance of lst Bn 116th Inf to POINTE DU HOE.

Upon arrival of lst Bn 116th Inf at POINTE DU HOE, Ranger Group is attached to lst Bn 116th Infantry.

(4) Ranger Group will maintain contact with CT and lst Bn 116th Inf CP by Liaison Officer with SCR 300s, and will submit half hourly reports by most rapid means available.

APPENDIX H

AFTER ACTION REPORT

MOTOR LAUNCH 304

The Commanding Officer,

H.M.M.L. 304,

c/o G.P.O.,

LONDON.

12th June 1944.

The Commanding Officer,

H.M.S. PRINCE CHARLES,

Commander, Assault Group O-4

Sir,

I have the honour to submit the following report of proceedings on the far shore on the morning of 6th June 1944.

At H-120 (0430B) I proceeded in accordance with previous

instructions and left the Transport Area with L.C.A. flotillas from L.S.I.'s "Ben my Chree" and "Amsterdam" formed up astern.

Course was set for Pointe du Hoe and this was checked by Q.H.2 and Radar Type 970. The D.R. position was not considered reliable owing to the slow speed of advance, the relatively strong effect of wind and tide and the difficulty of steering an accurate course in the sea which was running.

As the light increased and the coastline became more visible, the problem of identifying Point du Hoc arose. This promised to be difficult, since the coast was being bombarded heavily by the Gunfire Support Group and the headlands and cliffs were in many places obscured by thick clouds of smoke.

At this point, H-60, both Radar and the Q.H. appeared to be jammed. Subsequently this was found to be due to a faulty power supply, but from this time onwards they gave no assistance and it was necessary to rely on D.R. and visual recognition. It was thought that the heavy gun battery on Pointe du Hoe would disclose its own position, but the battery was not firing.

At about three miles from the headland which was thought to be Pointe du Hoc the problem of identification became especially difficult. The point was smothered in smoke and the bombardment was causing the cliffs to crumble and collapse in many places. The silhouette of Pointe du Hoe appeared different from the photographs supplied as a result of this bombardment, but a little further to the eastward was another headland which it resembled closely and course was modified to reach it.

At 4000 yards the L.C.T. was instructed to stop and the "Ducks" were launched.

At 1000 yards, when the L.C.A.'s were about to go in, it became apparent that this point was not Pointe du Hoe, since although the appearance coincided with the photographs, there were no concrete emplacements visible. Accordingly course was modified to reach the original point for which we had steered. As a result the landing was delayed until H + 30. No opposition was encountered throughout, save for some light machine-gun fire from the cliffs. These were engaged with Oerlikon fire and the L.C.S.s [LCS 91 and 102] also replied with .5" [M. G.].

The landing was effected without opposition and at this point the top of the cliff became clearly visible. The smoke had cleared away and several huts and concrete emplacements could be observed. While the landing was in progress, I engaged these targets with 3 Pdr. and 20-mm Oerlikon fire at ranges from 700–1000 yards. 20 rounds of 3 Pdr. F.S. were fired into the huts and emplacements and 1000 rounds of Oerlikon were used to spray the top of the cliff until the ascent was completed. A few figures were discerned moving about on the top of the cliff, but these were fired on and appeared to take cover.

Meanwhile an American destroyer came up and assisted in bombarding the cliff top, several hits being observed on the emplacements.

Some counter-fire was opened by a Bofors gun, but the firing was erratic and mostly astern or above us.

The L.C.A.'s were now leaving the beach. They had shipped a good deal of water owing to the heavy sea and there were several stragglers, for whom I waited. There was no further opposition from ashore and firing was ceased when the Rangers had gained the top of the cliff.

There appeared to be no underwater obstructions in the immediate vicinity of the point although a little further to the eastward and about 1000 yards off shore where I had gone to engage the light M.G. nests referred to earlier, a minefield was detonated immediately astern without causing us any damage.

When the L.C.A.'s had assembled, course was set for the transport area and I covered their rear until they were out of range of any possible batteries remaining.

It is infinitely regretted that our assistance to the Rangers was of such slight value, especially in the capacity for which we were specifically employed. While it is thought that Radar Type 970 together with Q.H. would have established definite identification, it is appreciated with great concern that our other resources proved inadequate in the presence of difficulties which it had been impossible to predict.

I have the honour to be, Sir,
Your obedient servant,

APPENDIX I

THINGS HISTORIANS DON'T KNOW

S ub-Lieutenant C. Beevor actually saved Omaha Beach through his navigational error. Because of his error, the 5th Ranger Infantry Battalion (RIB) was diverted to Omaha Dog Green Beach and saved the day at Omaha Dog Beach.

The 5th Ranger Infantry Battalion landed intact as a full-strength Ranger Infantry Battalion on Omaha Dog Red Beach because of several factors:

- Sub-Lieutenant Beevor's navigational error which forced Ranger Force C to divert to Omaha Dog Green Beach (Exit D-1).

- Ensign Victor Hicken's report to Dog Green Landing Control caused the closure of Dog Green Beach to landings just after 0700. As a result Ranger Force C was diverted by landing control to Dog White Beach.

- The fierce resistance met by the first wave of Ranger Force C (Group and Battalion Headquarters & A & B Companies, 2nd RIB) forced Lieutenant Colonel Schneider, the Force C commander, to divert the second and third waves of Ranger Force C (the 5th Ranger Infantry Battalion) from Dog White to Dog Red.

- Omaha Dog Red was fairly quiet because of several additional factors:

  1. WN 70 was knocked out by at least one of the following units that landed in front of it. One or two boat sections of B Company, 116th Infantry and/or B Company, 2nd Ranger Infantry Battalion with later help from A Company, 2nd Rangers.

  2. The brush fire on the face of the bluffs overlooking Dog Red masked the beach from vision of defenders above it and forced many defenders to withdraw from their prepared positions. The fire also exposed many of the emplaced anti-personnel mines.

  3. WN 69 could not observe or fire on Dog Red. This was due to both smoke and terrain, a knob that partially blocked the view. WN 69 was also under direct attack by elements of the 2nd Battalion, 116th Infantry led by Major Sidney Bingham. Attacks such as Bingham's prevented many WNs from undertaking their real mission of providing interlocking defensive fire to protect adjacent WNs

and the entrenched positions on the bluffs between the WNs.

4. The breakwaters (retards) formed protective walls between the 5th Rangers and the dreadful weapons fire that poured down the beach from the bluffs to the west, from WN 70 while it was still operating and the WNs at D-1.

5. And most of the artillery fire from the vicinity of D-1 was directed at the lucrative targets—landing craft coming in—and not at troops that had debouched and were scattered on the beach. Artillery fire from inland batteries had to clear the mask of the bluffs and was unable to hit on the beaches near the seawall. These indirect fire rounds hit out in the sea approaching the beach.

APPENDIX J
PHOTOGRAPHS

Captured German photograph purportedly of one of the six 155-mm guns in position at Pointe du Hoc.

Omaha Dog White Beach with edge of Dog Red on left, May 19, 1944

These are the breakwaters on Omaha Dog Red Beach. Raaen landed on the extreme right of the picture. Note the switchback path on the bluffs. Many used this path ascending the bluffs.

Omaha Beach as seen from Exit F1. Note Pointe et Raz de la Percée in the background.

The Blue and Yellow Ranger patch worn during the Invasion.

DC-666

Left: The eight DCS recipients from the 5th Rangers, from left: Max Schneider (CO), George P. Whittington (B Co.), Charles H. Parker (A Co.), Francis W. Dawson (D Co.), Willie Moody (C Co.), Howard McKissick (C Co.), Denzil R. Johnson (A Co.), Alexander W. Barber (HQ, Medic Det.). Not shown, Major Richard P. Sullivan (XO) and Lieutenant Joseph R. Lacy (Chaplain). These two received their awards from the Provisional Ranger Group.

Bottom: Officers of the 5th Rangers at Fort Dix, December 1943. Rear row: Askin, Runge, Wise, Eikner, Heffelfinger, Petrich, Garvik, Byrne, Carter, Butler, Murray, Raaen, Luther, Whittington, Suchier, Van Riper. Front row: Reville, Snyder, Gawler, Parker, Mehaffey, Marro, Miller, Dunegan, Pepper, Anderson.

Troops crowded in to an LCA.

HMS *Prince Baudouin* taking on LCAs, June 1944.

LCA 1377 taking troops to the *Baudouin* on June 1, 1944. Raaen is standing in front of the small signal mast. Major Sullivan stands to Raaen's left with Sub-Lieutenant Pallent between them.

Rangers embarking on June 1, 1944, for trip to the Landing Ships Infantry (LSI).

Colonel Rudder (left) and Chaplain Lacy congratulating each other after each was awarded the Distinguished Service Cross.

Raaen and Father Lacy, Normandy 1944

Raaen at Toul, November 1944

Today's and Yesterday's Rangers, then Captain Bill Linn and General
Raaen at the Colleville Cemetery, June 6, 2004.

INDEX

1st Infantry Division, 14, 34, 148

2nd Rangers, 4, 11, 16, 19-24, 26, 28, 32, 34, 36, 38, 45, 55, 58, 65, 77-80, 82, 84-85, 87-89, 91-92, 94-95, 100, 104-105, 109, 115, 132, 159

5th Rangers, 3, 7-8, 16, 19-20, 22, 24-25, 28, 33, 36-37, 39, 43, 46, 49-51, 54-55, 57, 66-68, 72-80, 82, 84-87, 90-92, 94-96, 98, 100-103, 105, 112-114, 118, 128-129, 131-134, 139-146, 149, 158-160, 165

   Company A, 3, 19, 37, 39, 48, 57, 59, 64-66, 72, 77, 95-98, 101-102, 104-108, 114, 117, 120-126, 125, 139, 165

   Company B, 3, 37, 40, 53-54, 57-58, 67-70, 72-73, 75-76, 79, 96, 98, 100-101, 121, 123-126, 139, 142, 165

   Company C, 19, 37, 43, 47-49, 53, 56, 58, 62-63, 75, 82, 87, 92, 94, 101-102, 105, 107-108, 116, 120-122, 140, 165

   Company D, 3, 37, 40, 47-48, 51-55, 57, 87, 90, 92, 101-102, 108, 116, 120-122, 125-127, 139, 165

   Company E, 3, 37, 40, 52-53, 57, 64, 72, 75, 96-98, 100-101, 116, 120-121, 123-127

   Company F, 3, 9, 13, 19, 37, 47, 59, 72, 78, 94-96, 98, 101-102, 105-108, 117, 119-120, 122-127

171

Headquarters Company, 3, 7, 16, 24, 37, 44, 46, 48, 53, 55-56, 58, 62, 64, 74, 80, 82, 101, 116-117, 132-138, 150

29th Infantry Division, 14, 20, 34, 39, 86, 91, 93, 108

58th Armored Field Artillery Battalion, 75, 86, 105, 126

116th Infantry Regiment, 16, 34-35, 44-45, 47, 54-57, 65-66, 68, 69, 72, 74-76, 79-80, 84-85, 87, 90-91, 96-98, 100-103, 115, 117, 121-126, 153, 159

165th Signal Photo Company, 4

293rd Joint Assault Signal Company, 4, 44, 133

501st Assault Flotilla, 11

504th Assault Flotilla, 10-11, 110, 114

507th Assault Flotilla, 10-11, 110, 113-114

701st Assault Flotilla, 36

743rd Tank Battalion, 21, 34-35, 46, 76, 80, 84-85, 87, 90, 96, 100-101, 121, 123, 125

Akers, Bernard, 54

Amsterdam, HMS, 25

Anderson, Lt. Dee, 52, 72

Arnold, Capt. Edgar L., 58

Askin, 1st Lt. Stanley, 9-10, 134, 165

Balkoski, Joe, 110, 115

Banning, Elmo E., 54

Beevor, Sub-Lt. Colin, RNVR, 24, 158

Bellows, Pfc. John, 105

Ben Machree, HMS, 25

Block, Dr. Capt. Walter E., 32

Brown, Lee, 85, 135

Burke, Jack, 105, 108, 117

Campbell, Robert M., 65

Canham, Col., 54, 69-70, 76, 79, 84, 87

Chateau de Vaumicel, 59, 64, 73, 76

Coughlin, Pvt. Francis T., 46, 135

Cota, BGen. Norman D., 45, 49-50, 52-54, 69, 75-76, 111, 116

Criqueville, 102

Dawson, Francis, 53, 121, 139, 143, 165

Dean, Minor, 85-86, 134

Dorman, Woody, 51, 53, 57

Dual Drive (DD) Tanks, 21, 34-36, 40, 46

Dunkle, Sergeant Henry J., 81, 85-86, 134

Easy Green Beach, 13, 34, 78

Eikner, Ike, 95, 165

Elsby, Kevan, 115

Englesqueville, 59, 77

Fabius exercise, 20

Force A, 16, 24, 26, 84

Force B, 20, 34, 82

Force C, 16-18, 20, 22-24, 33, 35, 71, 132, 158-159

Fox, William J., 14, 64

Gefosse-Fontenay, 106

Gelling, Lt. Cdr. W. E., RNR, 3, 10

Gerhardt, Gen. Charles, 86

Glasgow (British Cruiser), 32

Grandcamp-Les-Bains, 84, 100-101, 109, 125-127, 142

Graves, Sgt. James W. Jr., 7-8, 12, 36-37, 39, 50, 56, 60, 63, 71, 81, 110-113, 118, 135-136

Gregory, Lt., 69, 76

Gruchy, 88

Hamel au Pretre, 54, 66

Harwood, Capt. Jonathan H., 31-32

Hatfield, Thomas, 115

Hathaway, Sgt. Richard N., Jr., 48, 57, 137

Hawks, Capt. Berthier B., 45, 56, 85, 94

Hedgerows, 17, 54, 58-59, 60, 62-62, 67-70, 72-74, 77, 88, 95, 97, 102, 105, 109, 121

Helmet diamonds, 21, 30

Herring, Pfc. Thomas F. Herring, 49, 111, 116

H-Hour, 12, 21, 35, 152

Hicken, Ens. Victor, USNR, 35, 111, 115, 158

Isigny, 20, 84, 103, 109, 126-127

La Martiniere, 103-104, 106, 108

Lacy, 1st Lt. (Chaplain) Joseph R., 42-43, 128-131, 134, 139, 143-144, 168

Lambert, Pvt. Ingram E., 45

Landing Craft Assault (LCA), 6-7, 9-11, 13, 15, 26-27, 36-40, 42, 56, 59, 65-66, 78, 130, 132, 166-167

Landing Craft Personnel with Ramp (LCPR), 6

Landing Craft Vehicular Personnel (LCVP), 6, 95

LCA 578, 13

LCA 888, 26

LCA 1377, 6, 10-11, 132, 167

LCI 91, 40

LCI 92, 42

LCM, 39-40

LCT(A) 2227, 35, 111

Au Guay, 89-92

Les Moulins exit, 33-34, 66, 76

Les Perruques, 103-104, 106-108

Luther, Capt. Edward S., 40, 52-53, 72, 96-98, 111, 116, 165

MacDonald, 1st Sgt. Howard A., 116

McCullough, Pfc Artur J., 41, 43, 135

McIlwain, Walter (Mac), 54, 70, 111, 116

McKissick, Capt., 94, 99-100, 140, 146-147, 165

Maisy Battery, 102-109, 117

Merrill, Capt. Richard P., 36, 111, 115

Metcalf, Lt. Col., 76, 84, 87-88, 90-92, 104

Miller, 1st Lt. George, 47, 165

Model 209 Code Converter, 8

Moody, Sgt., Willie, 94, 99-100, 140, 145-146, 165

Moore, Lt. Woodford O., 52, 64, 135

Omaha Beach

    Omaha Dog Beach, 14, 20, 33, 71, 143, 146, 158

    Omaha Dog Green, 20-21, 23, 33-36, 76, 120, 132, 158

        Easy Green Beach, 13, 34, 78

    Omaha Dog Red, 33, 37, 46, 130, 140, 142, 158-159, 162

    Omaha Dog White, 23, 33, 37-38, 44, 46, 66, 120, 132, 159, 162

Ormel Farm, 59, 64, 66, 73

Osmanville, 109, 126, 127

Pallent, Sub-Lt. Ernest, RNVR, 12, 167

Parker, 1st Lt. Charles H. (Ace), 39, 48, 59, 64, 66, 77-78, 106, 108, 110-112, 114, 139, 144, 165

Pepper, Lt. Bernard M., 67-69, 165

Petrick, Dr. Thomas G., 107, 135

Pingenot, Lt. Leo A., 65

Pogue, Sgt. Forrest, 114

Pointe du Hoc (Hoe), 15-17, 19-20, 22-27, 27, 69, 75, 77, 79, 84, 86-90, 94-96, 99-101, 104, 115, 122-126, 132, 141, 144-145, 152-153, 155-156, 161

Pointe et Raz de La Percée, 19-20, 23-24, 27, 46, 50, 65, 82, 163

Port-en-Bessin, 14

Prince Baudouin, HMS, 3-4, 7-8, 10-11, 13, 82, 110, 113, 119, 129-130, 132, 166-167

Prince Leopold, HMS, 3, 7, 113, 119, 132

Provisional Ranger Group, 4, 11, 16, 25, 165

Rahmlow, 1st Lt. John L., 47

Ranger Company organization, 18-19

Reardon, Lt. Col. Mark, 113

Reed, Bill, 51, 53, 57, 111, 116

Reville, 1st Lt. John J., 9, 13, 95, 107, 112, 117, 165

Rudder, Lt. Col. James E., 4, 16, 19, 22, 24-31, 77, 95, 99-100, 115, 168

Runge, 1st Lt. William M., 13, 95, 165

St. Laurent-Vierville, 13, 64, 76, 94, 119, 121-122

St. Pierre Du Mont, 88-93, 98-100, 104, 123-124, 151

Salomon, Lt., 82

Schneider, Lt. Col. Max F., 8, 10-11, 22, 24, 33, 36-40, 44, 47-50, 52, 54, 57, 63, 67, 71, 73-75, 79-81, 84, 87, 90, 94, 96, 100, 103, 108, 120, 132, 134, 139-141, 147, 159, 165

SCR 284 radio, 8

SCR 300 radio, 23, 25, 94, 153

Sharp, Cpl. Jack, 85-88, 135

Shea, 1st Lt. Jack, 50, 116

Shoebrick, USS, 103

Soper, Cpl. James F., 43, 135

Sterne, Gary, 112, 117

Stowe, Bob, 55

Street, Maj. Jack B., 95, 124

Stump, Pfc. William A., 49

Suchier, 1st Lt. Oscar A., 48, 56, 59, 165

Sullivan, Maj. Richard P., 3-4, 9-12, 16, 36, 41, 44, 47-48, 63, 70, 74, 85, 87-91, 101-102, 105, 107-109, 128-129, 132, 134, 139, 141-142, 147, 165, 167

Swartz, Lt. Stanley M., 45

Taylor, Ltd. Col. Charles, 89, 114-115

Taylor, Walter, 66

Teller mine, 40, 119

Texas, USS, 10, 12-13, 29, 32, 81, 95, 110, 113-114

Van Riper, Howard, 56, 62, 83, 134, 165

Vierville exit, 16, 18, 20, 33-35, 44, 50, 65, 76, 86, 96

Vierville-Grandcamp Coastal Road, 20, 77

Vierville-sur-Mer, 20, 71, 78, 121-124, 141, 150

Vire Estuary, 14

Vire River Valley/Vire River, 103-104, 127

Vulle, Tony, 56

Wells, T-5 Theodore H., 19, 22, 24-32, 111, 115, 132, 134

West, Lt. E. H., 10

Whittington, George P., 40, 53-55, 67-69, 139, 142-143, 165

Widerstandsnester (WN), 17-18, 36, 65, 104, 159-160

Williams, Lt. William B., 65-66

Wise, Capt. Wilmer K. (Bill), 4, 47, 63, 93-94, 165

Zelepsky, Lt., 78